BRENT BUTTERWORTH

PUBLICATIONS INTERNATIONAL, LTD.

The brand name products mentioned or shown in this publication are service marks or trademarks of their respective companies. Mention of these products in text, photographs, or directions is for demonstration purposes only and does not constitute an endorsement of Publications International, Ltd., nor does this mention imply that these products are any more or less likely to fail than any other products. The use of electrical equipment is dangerous, and any repairs should be attempted with caution.

Brent Butterworth is editor of *Home Theater* magazine and former senior editor of *Video* magazine. He has written dozens of articles on VCR technology and maintenance, and has produced and engineered commercial and artistic video projects since the mid-'80s. He has also written technical articles for magazines such as *Omni, Bicycling, Wired,* and *Audio,* and Consumer Guide®'s *The Big Book of How Things Work.*

Photography: Sam Griffith

Illustrator: Gene Givan

Model: Rick Welch/Royal Model Agency

Special acknowledgment to Jack Melnick and Noor Bhakrani of Melnick Service Works, Deerfield, Illinois, for technical assistance and consultation in the preparation of this book.

CONTENTS

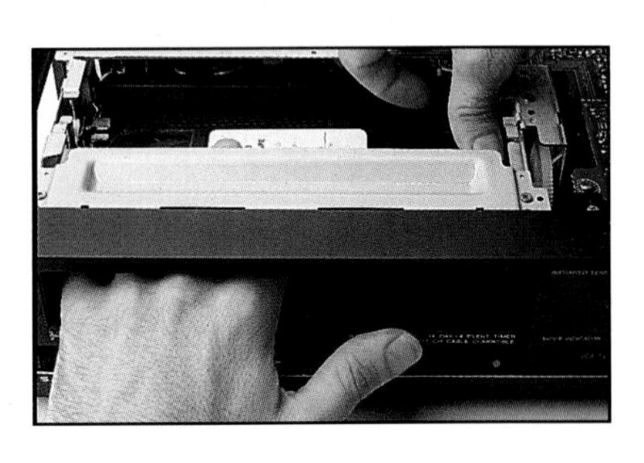

THE VCR: A SUCCESS STORY

Few high-tech products have affected our lives as much as the VCR (videocassette recorder). Its capabilities—recording and playing back tapes of movies and TV programs—seduced the world so quickly that it became one of the biggest success stories in the history of manufacturing. By 1995, VCRs had found their way into nearly 80 percent of U.S. homes.

The first camcorders were quite large, heavy, and difficult to handle.

Thanks to the VCR, people can now watch what they want when they want. Anyone can stroll down to the corner drugstore and rent an Oscar-winning blockbuster any night of the week. Any morning offers the opportunity to exercise with celebrities and famous athletes. And of the myriad hobbies and sports, tapes are available that offer tips from top pros.

But 20 years ago, the VCR couldn't even get its foot in the door of American homes. Many companies tried introducing videotape recorders in now forgotten formats such as V-cord, Technicolor, and Cartrivision, all without success.

One of the formats introduced for home use in the early 1970s was Sony's ¾-inch U-matic format. The gigantic machine was a flop on the consumer market, but professional videographers and schools and other institutions loved it. Sony used the experience gained in the production of U-matic VCRs to produce the first successful home deck, the Betamax, in 1975.

Hindered by high prices and a maximum recording time of just one hour, the Betamax didn't catch on like wildfire. But a competing format from JVC, the VHS, did. Introduced in 1977 under the RCA brand, the first VHS deck offered two hours of recording time at the standard SP speed and four hours of recording time at the long-play LP speed. Sony eventually boosted the recording times of its Beta decks, but Beta never caught up to VHS. Despite numerous technological firsts—stereo hi-fi sound and the SuperBeta format with enhanced resolution—Beta machines had all but faded away by the late 1980s.

But home video technology has not stood still. Many new formats and capabilities have been introduced since the mid-1980s. The first of these was the camcorder, which combined a video camera with a VCR in one hand-held unit. Next was the 8mm format, a tiny cassette designed for camcorders by a consortium of VCR manufacturers. The camcorders for 8mm are much smaller than earlier camcorders and are easier to hold and handle.

High-resolution versions of each major format have also been introduced, bringing the quality of professional video equipment into consumers' hands. Now, not only can consumers watch any program they want at any time, but they can also make their own programs using affordable home video equipment. Consumer versions of professional special effects devices have even been introduced. In fact,

As technology improved, new and easier-to-handle formats (such as the 8mm format shown here) made smaller camcorders possible.

Early VHS VCRs, such as this model from the late 1970s, opened up the VCR market to the consumer.

the best home videos are indistinguishable from professional videos.

How to Use This Book

This publication will provide both experienced electronics buffs and those totally confused by electronic products with useful information that will help solve most problems encountered with a VCR. The book need not be read all the way through; time can be saved by turning straight to the section that deals with the problem encountered.

Questions about choosing a VCR are dealt with in Chapter 1, "VCR Formats and Features." Specific questions about hooking up and using a VCR are addressed in Chapter 2, "How to Hook Up and Use a VCR." To get a quick education on the mechanics of a VCR, turn to Chapter 3, "How a VCR Works." Chapter 4, "The Care and Maintenance of a VCR," explains how to keep a VCR running at its best. And Chapter 5, "Troubleshooting," provides a basic guide to fixing simple problems that often occur with VCRs. At the back, a glossary defines some simple terms, and an index can help locate information that can't be found right away.

But once the immediate problem has been solved, keep reading. There's plenty of information about features that you might want on a new VCR, and a VCR already hooked up in the home may have features that you have not explored yet. This book should help you avoid expensive repair bills and leave frustration behind.

VCR FORMATS AND FEATURES

VCR Formats

Currently, consumers can choose from seven videotape formats. Whether that's a good or bad situation depends on an individual's point of view. This large selection makes it possible to get exactly the capabilities and performance required, but it also makes choosing a format much more confusing.

By far, the most popular format for VCRs is VHS. Introduced in 1977, VHS became popular because it offered long recording times. Even the earliest VHS VCRs provided four hours of recording time, and the latest T-200 tapes can hold up to ten hours of programming.

VHS uses ½-inch tape wound into a cassette measuring about 7½ inches long by 4 inches wide by 1 inch thick. It produces about 240 lines of horizontal resolution. (*Horizontal resolution* is the number of lines produced across the screen—the more lines, the sharper the picture. Television broadcasts have about 330 lines of horizontal resolution.) VHS VCRs are available in inexpensive monophonic versions, owned by millions of people around the world, as well as in hi-fi stereo versions that provide much better sound. There's also a smaller version for camcorders called VHS Compact, or VHS-C. VHS-C cassettes fit into an adapter

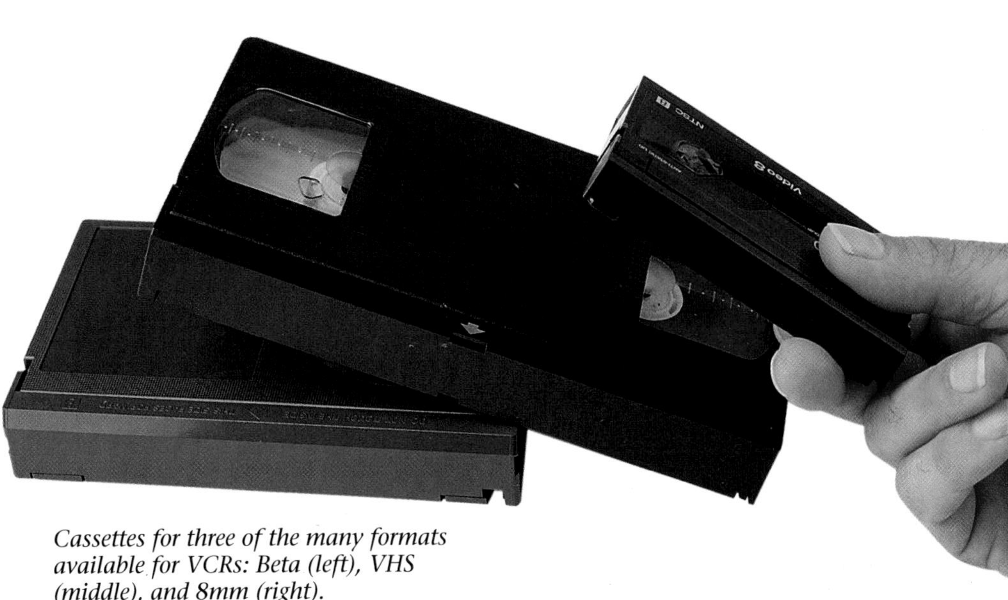

Cassettes for three of the many formats available for VCRs: Beta (left), VHS (middle), and 8mm (right).

so that they can be played on a VHS VCR.

Beta—the first home video format to gain wide acceptance—was introduced in 1975. However, it has been almost completely pushed out of the consumer market—in 1995, only Sony still sold Beta decks, and they offered only two models. Beta records up to five hours on ½-inch tape wound into a cassette slightly smaller than a VHS cassette. Like VHS, Beta is available in monophonic and stereo hi-fi models. The original Beta format produced about 240 lines of horizontal resolution, but a later version—SuperBeta—produces about 300 lines. There is an even more improved version called Extended Definition Beta (ED Beta), which produces about 500 lines of horizontal resolution. However, ED Beta VCRs and tapes have never become very popular.

VHS has also been upgraded for a better picture. The Super VHS (S-VHS) format, introduced in 1987, produces about 400 lines of horizontal resolution using a special version of VHS tape. All S-VHS VCRs also play and record regular VHS tapes. Super VHS Compact, or S-VHS-C, is available for camcorders, and like VHS-C tapes, S-VHS-C tapes can be placed in an adapter for playback in an S-VHS VCR. S-VHS VCRs use a special connector, called an S-video connector, to keep the brightness and color signals separate. This helps prevent hanging dots, which are the "crawling" horizontal edges that may be noticed in some television pictures. Many newer TV sets have S-video jacks for use with S-VHS VCRs. Even if a TV set doesn't have such jacks, S-VHS still provides a better picture.

Designed for camcorders, the tiny 8mm format is also available in VCRs. All 8mm VCRs have hi-fi sound (either mono or stereo), and some even have stereo digital sound similar to that produced by CDs. This format produces about 250 lines of horizontal resolution. Hi8, an upgraded version of 8mm, produces about 400 lines. The tapes fit easily into a shirt pocket.

Features to Expect on a VCR

The earliest VCRs didn't have many features, but video technology has come so far since then that consumers can ex-

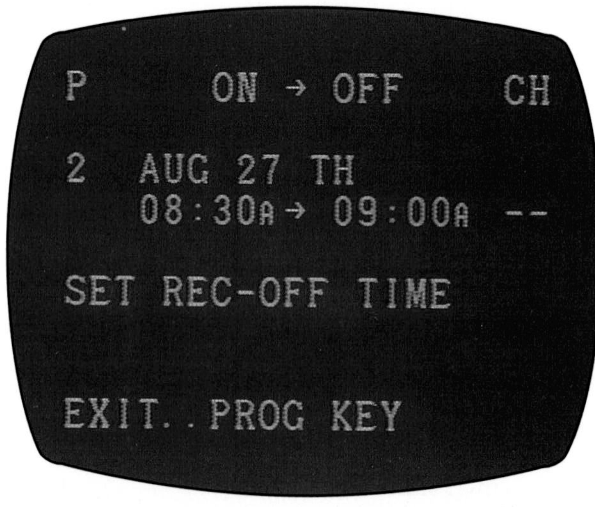

A typical programming menu as seen on a TV screen. This feature can be expected on most of today's VCRs.

pect a wide range of advanced features on a new VCR.

One feature that has become standard is *a wireless remote control.* In fact, many of today's VCRs have almost no controls on the deck—they rely almost totally on the remote. Wireless remote controls—once found only on the most expensive decks—now come with even the least expensive VCRs.

A real-time counter (box), which is more useful than the older mechanical counters, is standard on today's VCRs.

Home VCRs have always come with timers to record TV programs while no one is at home. But it has become almost a joke that no one actually knows how to use a VCR timer. Older VCRs were very difficult to program, but almost all newer decks come with easy-to-use on-screen programming systems.

On-screen programming, usually activated from the remote, employs graphic menus on the TV screen to guide the user through the programming process. The VCR asks what day, time, and channel to record and for how long. Most people find this system much easier to use than the tiny front-panel buttons used on early VCRs.

All current VHS VCRs use an improved recording system called *HQ,* which stands for "high quality." HQ comprises several small improvements designed to enhance picture quality. These improvements include video noise reduction, which helps eliminate the white specks and streaks (snow) that may be seen on poor-quality tapes.

Three recording speeds are available in VHS: the two-hour *standard SP speed,* the four-hour *long-play LP speed,* and the six-hour *extended-play EP speed* (also called SLP). These speeds refer to the maximum recording time when using a standard T-120 tape. Generally, the longer the recording time, the lower the picture quality.

All but the earliest VHS VCRs play all three speeds, and most record in all three speeds. However, some of the latest decks don't record in LP speed. VCR manufacturers have improved the quality of the EP speed significantly and don't feel the LP speed is necessary anymore. Some people, though, prefer to have the option of recording in LP speed.

A real-time counter shows in hours, minutes, and seconds how much of the tape has played. This type of counter is much easier to use than the counters on early VCRs, which simply supplied a number that didn't correspond to a specific segment of time. Real-time counters are especially useful for recording TV programs. For example, if several sitcoms are recorded on one tape, the starting point of each show can be quickly found by fast-forwarding through 30 minute's worth of tape.

A VCR with two video heads gives a snowy picture when in pause (top), while a VCR with four heads gives a much clearer picture in pause (bottom).

Nearly all newer VCRs allow channel selection using a numerical keypad or by flipping through the channels with the up and down buttons. Most VCRs also have *channel programming* to bypass channels that either have no programming or have programs of no interest to the viewer. For example, a cable user may receive, but never watch, channels 3, 6, 8, and 10 (or someone using an antenna may have no programming on those channels). The VCR can be set up so those channels come up only if they are punched into the keypad. Then, when using the up and down buttons, the channel will go directly from 2 to 4. Most VCRs also have *automatic channel programming,* which scans through all channels and automatically deletes any empty ones.

Features to Consider

While today's VCRs come loaded with standard features, the growth in video technology has given rise to many advanced features. For renting movies and occasionally recording a TV show, many of these special features may not be needed or wanted. However, one or more of these features may solve a specific problem or make operating a VCR much easier.

The most popular special feature is *four heads*. The use of four video heads does little to improve a VCR's picture in the normal play mode, but it does wonders with the VCR in pause. Two-head VCRs produce a noisy picture in pause, with white streaks obscuring most of the details. But four-head VCRs produce clean still frames, with little or no noise. They also produce cleaner pictures in scan and slow motion modes. Four heads are best for pausing to examine the picture more carefully or to scan through a tape to find a particular spot.

While on-screen programming has made it much easier to program a VCR, manufacturers are still coming up with new methods that promise to make the procedure almost automatic. The most popular of these is *VCR Plus*. To program a VCR equipped with VCR Plus circuitry, simply punch in a four- to seven-digit code number, and the VCR figures out the rest. VCR Plus codes can be found in many major newspapers and in *TV Guide.* Some VCRs equipped with VCR Plus even have an infrared emitter that can change channels on your cable box, so you don't have to remember to set the channel on the box when you program the VCR.

A few new VCRs feature *automatic clock setting.* They use a time and date signal transmitted by some TV stations, so you never again have to face a blinking "12:00."

A new service called *StarSight,* also transmitted by TV stations, combines automatic programming with an on-screen channel guide. Using StarSight, you can browse through the broadcast and cable TV stations. If you see a listing for a show you want to record, just highlight that listing and the VCR automatically records the show.

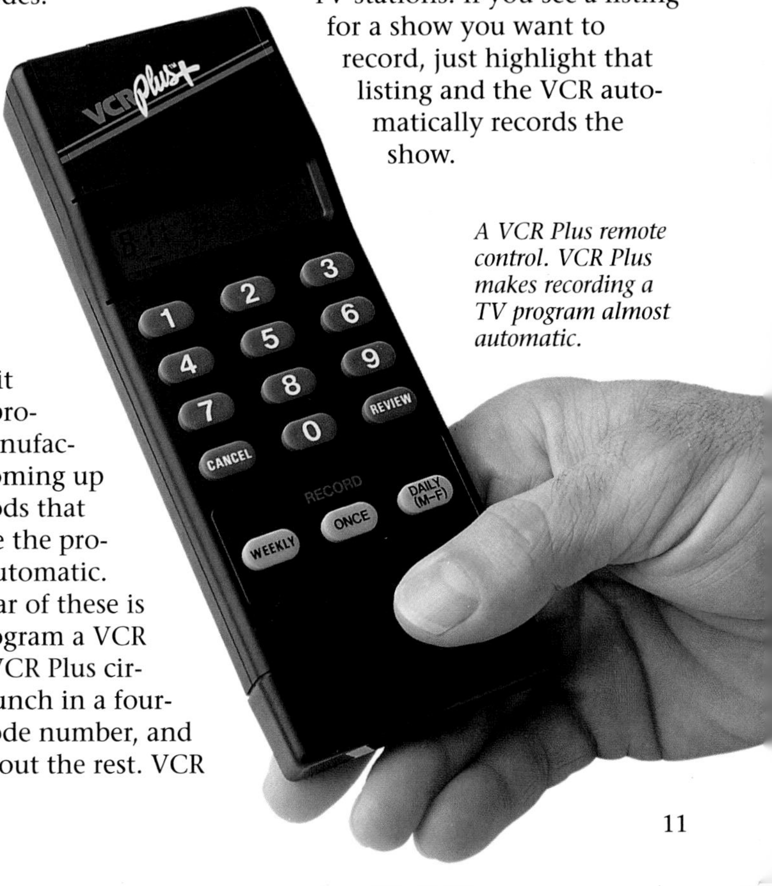

A VCR Plus remote control. VCR Plus makes recording a TV program almost automatic.

VCR Plus is also available as a separate remote control that works with almost any VCR. Most people feel VCR Plus is the easiest way to program a VCR. But before buying a VCR Plus deck or remote, make sure the local paper carries the codes.

Another new programming system is the *LCD Program Director,* which is found on many Panasonic VCRs. It uses dedicated controls—either on the remote, on the VCR, or on a special Program Director remote—to program the VCR. Sony uses a similar system on some of its VCRs. Both systems are almost as easy to use as VCR Plus and don't require special codes.

Indexing allows electronic marking of specific spots on a tape to automatically find that spot later. Some VCRs mark index points automatically at the spot on the tape where recording starts. Other VCRs allow index points to be placed wherever the user wishes. In both systems, either a dedicated button or an on-screen menu system activates the index search feature, and the VCR automatically fast-forwards to the next index point.

Almost all newer VCRs now do some of the work for the consumer. Most decks automatically play any tape that has its record safety tab removed and then rewind the tape when it reaches the end. Some VCRs will then eject the tape and shut themselves off, which is ideal for those who like to fall asleep with the TV on. Some VCRs automatically rewind a tape when they hit a blank spot on the tape. On the most advanced VCRs, these automatic control sequences can be deactivated if desired.

Automatic head cleaners make VCRs almost maintenance-free. A tiny cleaning roller automatically rubs against the video head drum every time a tape is inserted. This means the heads of a VCR may never have to be cleaned. However, the transport mechanism of a VCR may still need cleaning from time to time.

Children seem to have a natural curiosity about electronic products, and they often accidentally damage the products while indulging their curiosity. With VCRs, this damage usually comes in the form of objects stuffed into the tape hatch. Many youngsters have even tried to feed the VCR by shoving food down the hatch. Since any foreign object in the tape hatch can easily damage or destroy a VCR, some manufacturers have added a *safety latch* that locks the tape hatch in place. The latch is usually hidden behind a door.

Features Designed for Camcorder Owners

As camcorders have become popular, manufacturers have introduced VCRs designed for editing camcorder videos. Editing cuts out the unwanted parts of a video while keeping the desired parts. Sometimes, titles and special effects can be added during the process.

The most important of these editing features is the *flying erase head.* A flying erase head rides

The jog/shuttle dial (box) is especially useful for those who wish to edit tapes made with a camcorder.

along on the head drum and cleanly erases video tracks so glitches or rainbow patterns don't appear between scenes.

Front-panel VCR inputs make plugging a camcorder into the deck easier because no one has to reach around the back to change cables. This is an especially convenient feature for those who own 8mm or Hi8 camcorders, since those tapes can't be played on VHS VCRs.

To easily find the exact starting and ending points of a scene, a *jog/shuttle dial* will do the trick. This control comprises a shuttle ring that surrounds a jog dial. The shuttle ring works like the fast forward and rewind controls, while the jog dial allows stepping forward or backward through a tape frame by frame.

An *audio dub* allows the addition of narration, background music, or sound effects to a tape while keeping the original video. A *video dub* does the opposite—adding new video over existing audio. Both are extremely useful for creative editing, but neither 8mm nor Hi8 VCRs have video dub.

Insert edit, which is available only on VCRs with flying erase heads, lets the editor insert a new scene between two existing scenes on a tape without leaving glitches at the beginning or end of the scene.

A *synchro edit jack* makes dubbing and editing a video very quick and easy: Just punch one button to start or stop two machines at the same time. This feature usually works only if both machines equipped with the synchro edit feature (for example, a camcorder and a VCR, or two VCRs) are the same brand.

To jazz up a video, use *digital effects.* Many effects are available, including strobe (seen in many music videos), paint (which converts the picture into a colorful, splotchy scene resembling a watercolor painting), and mosaic (which converts a picture into a series of tiny squares).

A *character generator* lets the user electronically add text to a video, as seen in TV sports and newscasts. Most offer eight colors, and some provide up to four text sizes.

Features for Use in Home Theater

The latest trend in audio/video is home theater. Home theater systems combine high-quality surround sound with large-screen video, providing an experience that's almost like being at a movie house.

A *hi-fi stereo* VCR is required to take advantage of a home theater setup. VHS hi-fi VCRs record audio signals under the video signals, using audio heads that ride on the head drum with the video heads. The sound from a hi-fi VCR is better than that of most audio cassette decks. In fact, many people use hi-fi VCRs to record radio programs or party tapes.

Most movies available for rental have hi-fi stereo tracks. The speakers of a stereo system will offer much better sound than a TV's speakers; simply connect the hi-fi VCR to the stereo system.

Many movies also have surround sound tracks embedded in the stereo signal. These tracks can be extracted with a *surround sound decoder* or an *audio/video receiver.* This provides right and left stereo signals as well as a center channel signal and a rear speaker signal. But without a hi-fi stereo VCR, there is no way to take advantage of tapes encoded with surround sound.

One drawback of a complex audio/video system is the clutter of remote controls that ends up on the coffee table. Some VCRs solve this problem with *universal remote controls*. These control the VCR,

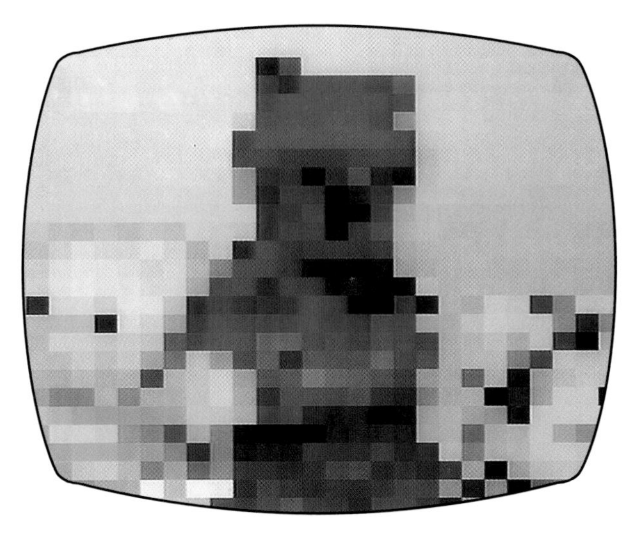

These pictures show various digital effects. The original view is shown at top left. Strobe (top right), paint (bottom left), and mosaic (bottom right) can all be used to make a video more interesting.

the TV set, and sometimes other components as well.

Many VCRs have TV controls that work only if the VCR and TV are the same brand. Others use *preprogrammed code sets:* Just punch in a two-digit brand code listed in the owner's manual, and the VCR remote is configured to operate the TV. Some VCR remotes also carry codes for cable boxes. And some VCR remote controls can learn commands from other wireless remote controls, so they can be used with any piece of equipment desired.

HOW TO HOOK UP AND USE A VCR

For most people, a video system consists of only a television set and a VCR. But even a simple two-piece system can be difficult to set up. When a cable box is included, many people may never figure out the correct combination. Fortunately, there's no real mystery to hooking up a VCR. Electronic devices work logically, and by thinking logically, anyone can easily master their use.

The above picture shows audio/video jacks (box) on the back of a TV. Many TVs do not have these jacks.

Hookup Using an Antenna

If the TV uses an antenna and has no audio/video jacks, hooking up the VCR is usually just a matter of connecting two wires and the VCR's power cord. Plug in the power cord first. Now connect the VCR and the TV with a coaxial cable. Most VCRs come with a coaxial cable, and electronics stores carry replacements. Connect this cable between the threaded F-connector on the VCR labeled "out to TV" and the matching threaded F-connector on the back of the TV set, usually labeled "antenna" or "ant/cable."

If the TV set has screws for the antenna connection instead of a threaded F-connector, a device

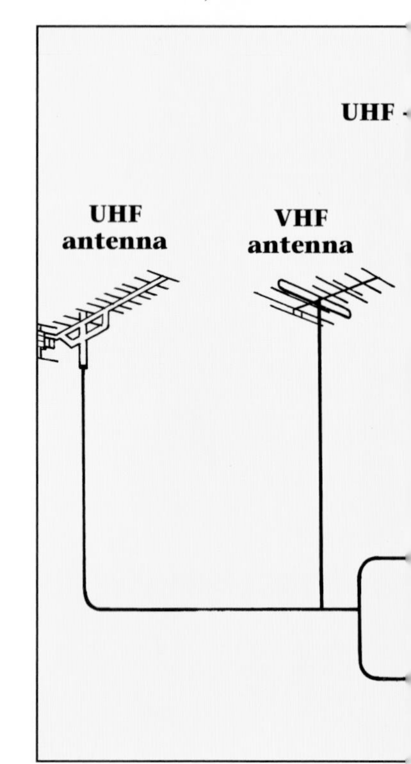

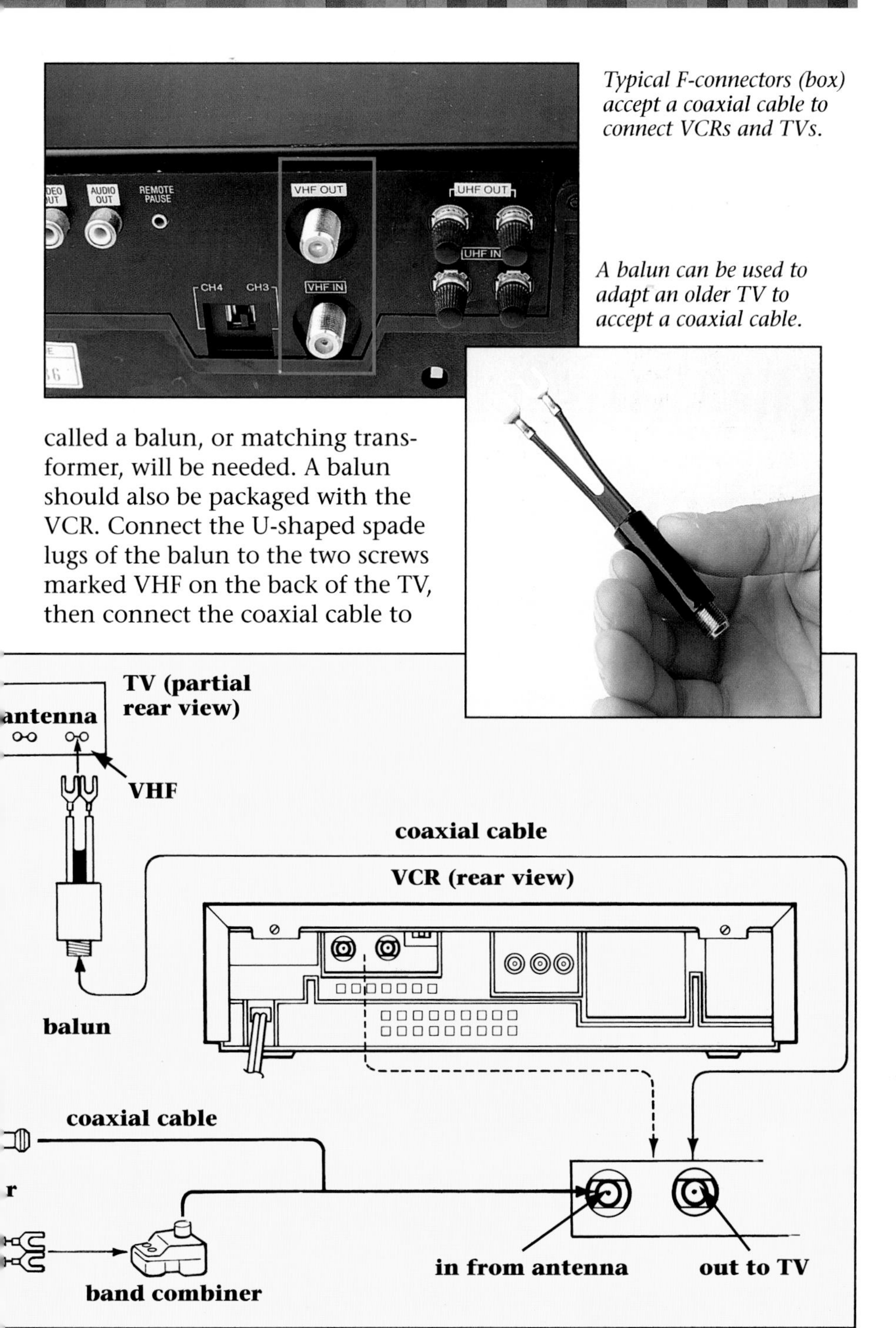

Typical F-connectors (box) accept a coaxial cable to connect VCRs and TVs.

A balun can be used to adapt an older TV to accept a coaxial cable.

called a balun, or matching transformer, will be needed. A balun should also be packaged with the VCR. Connect the U-shaped spade lugs of the balun to the two screws marked VHF on the back of the TV, then connect the coaxial cable to

This schematic illustration shows the basic hookup of an outside antenna, a VCR, and a TV.

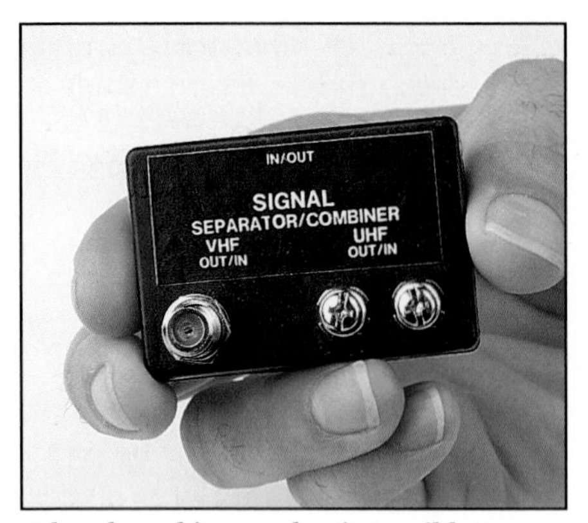

A band combiner makes it possible to connect an older outside antenna to a VCR via a coaxial cable.

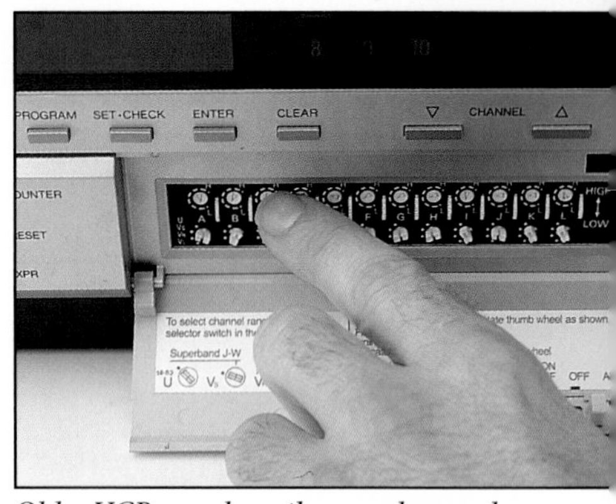

Older VCRs, such as the one shown above, use thumbwheels to tune in individual stations.

the threaded F-connector on the balun.

Finally, connect the regular TV antenna to the VCR. A rooftop antenna often has a single cable that carries both UHF and VHF signals. In this case, simply plug this cable into the F-connector on the back of the VCR that is labeled "in from antenna." If the antenna has separate cables for VHF and UHF, connect these cables to a band combiner (available at electronics stores) and then connect the band combiner to the VCR. Some older VCRs have separate screw terminal inputs for VHF and UHF, making a band combiner unnecessary.

After these cables have been connected, set the switch on the back of the VCR to channel 3 or 4 (whichever is unused in the local area). If the VCR has a "cable/antenna" or "CATV/normal" setting (activated through a switch or menu system), use the "normal" or "antenna" setting.

Newer VCRs have tuners that are permanently preset to standard TV channels. But some older VCRs require tuning in individual channels with thumbwheels. Each

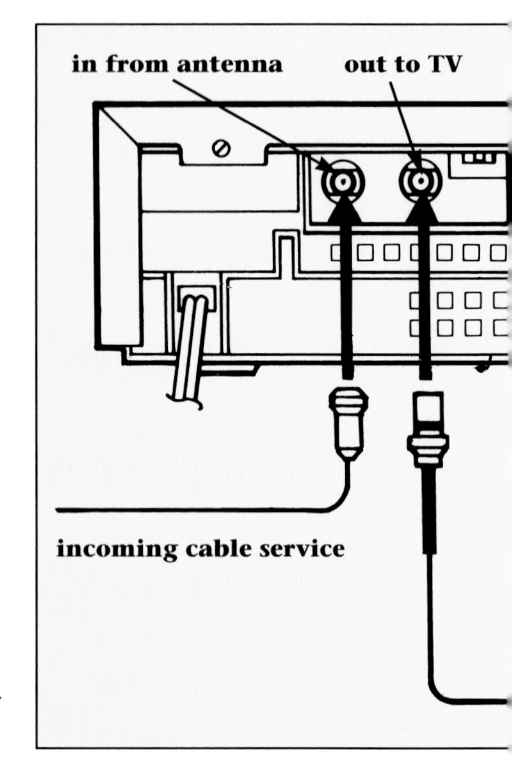

thumbwheel has a range selector switch, with positions for VHF low (channels 2 to 6), VHF high (channels 7 to 13), and UHF (channels 14 and higher). Set each channel by selecting the range (VHF low, VHF high, or UHF) and then tune in the channel by turning the thumbwheel. To make this easier, set the "VCR/TV" switch on the VCR to "TV" and tune in the desired channel on the TV. Make note of the program that's playing, then switch back to the "VCR" setting on the "VCR/TV" switch and search for the program with the thumbwheel.

If the VCR has auto channel programming, activate it. This feature searches through the channels and eliminates the inactive ones from the scanning sequence. This eliminates static when flipping through channels.

Hookup for Cable TV

Before trying to connect a VCR to a cable TV system, some basic information about the cable system is needed. If the system is only basic cable with no pay channels and only the pay channels are scrambled, a cable box probably isn't needed. If the user subscribes to pay channels, a cable box is necessary for only those channels. If every channel on the system is scrambled, a cable box

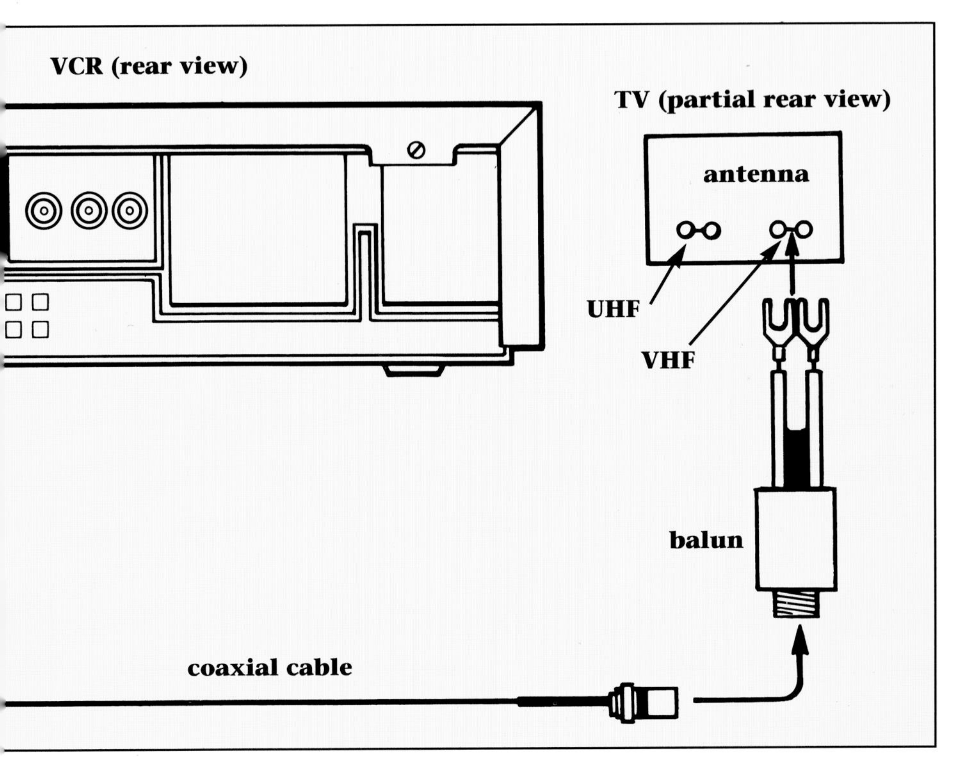

This schematic illustration shows the basic VCR and TV hookup for a cable service without a cable box.

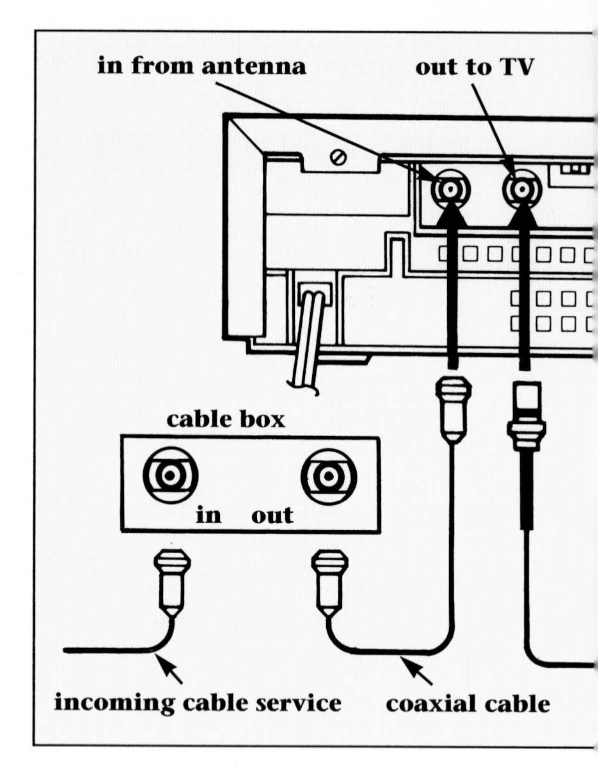

must be used all the time, and the viewer won't be able to use some of the VCR's features.

With no cable box, hooking up the VCR is simple. If the TV has no audio/video jacks, just connect two wires and the VCR's power cord. Plug in the VCR's power cord first. Now connect the VCR and the TV with a coaxial cable. Most VCRs come with a coaxial cable, and electronics stores carry replacements. The cable installer may also have one to spare. Connect this cable between the threaded F-connector labeled "out to TV" and the matching threaded connector on the back of your TV set, usually labeled "antenna" or "ant/cable."

If the TV set has screws for the antenna connection instead of a threaded F-connector, the cable installer should have fitted it with a device called a balun, or matching transformer. Disconnect the cable from the balun and plug the coaxial cable coming from the VCR into the balun.

Finally, connect the cable TV system to the VCR. Simply plug this cable into the F-connector on the back of the VCR that is labeled "in from antenna."

After connecting these cables, set the switch on the VCR to channel 3 or 4 (whichever is unused in the local area). If the VCR is cable-ready, set the "cable/antenna" or "CATV/normal" switch to "cable" or "CATV." Sometimes this switch is built into the VCR's menu system.

Newer VCRs have tuners that are permanently preset to standard TV channels. But some older VCRs require tuning in individual channels with thumbwheels. Each thumbwheel has a range selector switch, with positions for VHF low (channels 2 to 6), VHF high (channels 7 to 13), and UHF (channels 14 and higher). (If the VCR has thumbwheels, a cable box is needed to get channels 14 and higher.) Set each channel by selecting the range (VHF low, VHF high, or UHF) and tuning in the channel by turning the thumbwheel. To make this easier, set the "VCR/TV" switch on the VCR to "TV" and tune in the desired channel on the

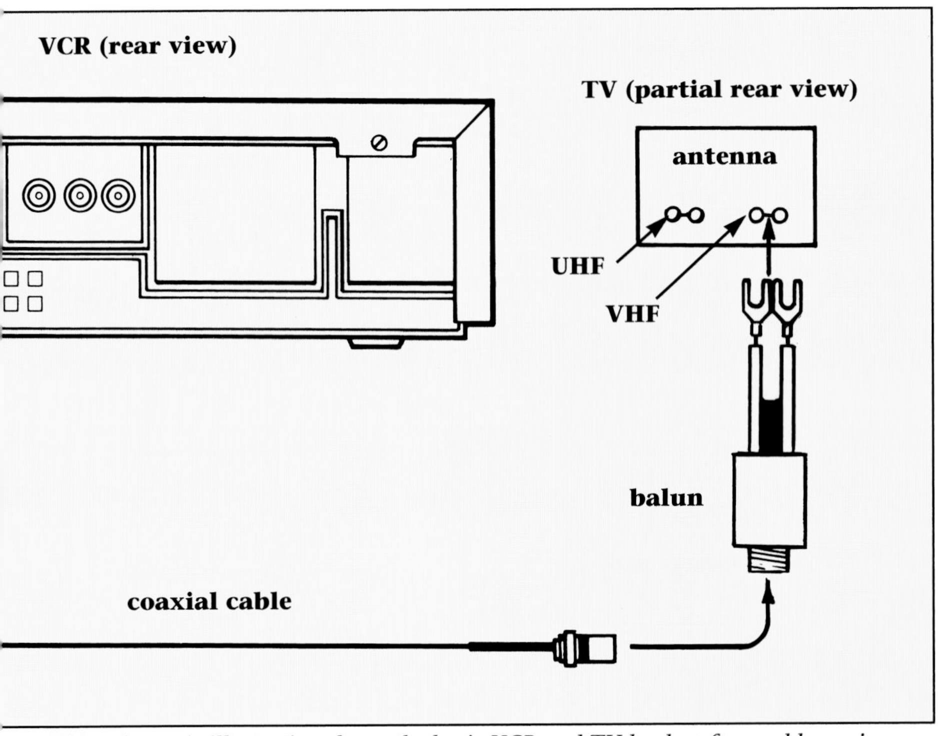

This schematic illustration shows the basic VCR and TV hookup for a cable service with a cable box.

TV. Make note of the program that's playing, then switch back to the "VCR" setting on the "VCR/TV" switch and search for the program.

For photos of audio/video inputs, a threaded F-connector, a balun, and thumbwheels, refer to the earlier section, "Hookup Using an Antenna."

Hookup with a Cable Box

If the user subscribes to pay channels, if all the channels on the cable system are scrambled, or if the VCR isn't cable-ready, a cable box must be used. This means operating the video system will be more difficult. To record a program from a pay channel, tune it in on the cable box first. To watch one channel while recording another, use a cable splitter and an A/B switch, which are available from the cable company or a local electronics store.

First, plug in the VCR. If the TV has no audio/video jacks, connect the VCR and the TV with a coaxial cable. Most VCRs come with a coaxial cable, and most electronics stores will carry replacements. The cable installer may also have one to spare. Connect this cable between the threaded F-connector on the VCR labeled "out to TV" and the matching threaded connector on the back of the TV

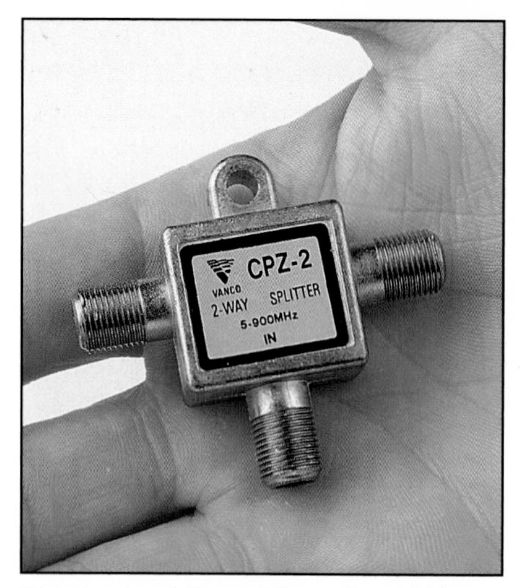

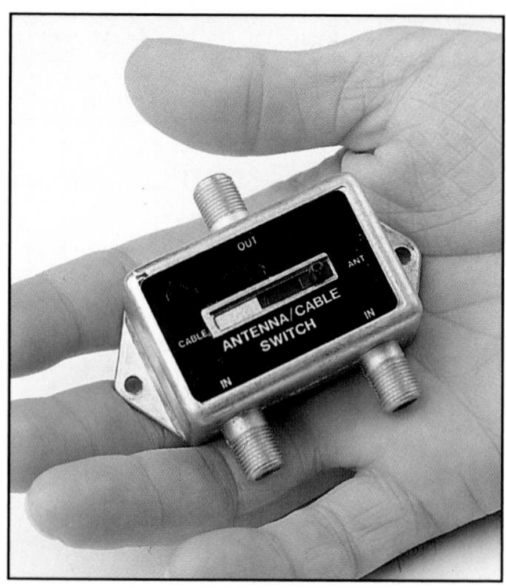

To get around the problem of watching one channel on cable while recording another channel, use a cable splitter (left) and an A/B switch (right).

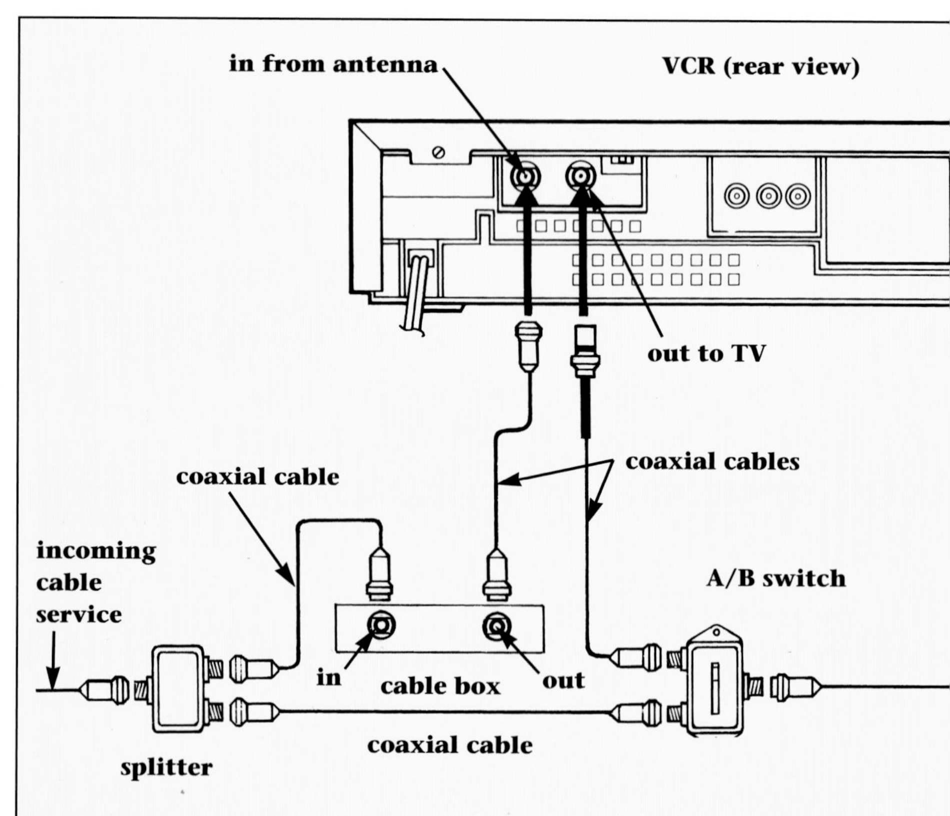

This schematic illustration shows how to hook up a cable system to a VCR and a TV so one program can be watched while another is recorded.

set, usually labeled "antenna" or "ant/cable."

If the TV set has screws for the antenna connection instead of a threaded F-connector, the cable installer should have fitted it with a device called a balun, or matching transformer. Disconnect the cable from the balun and plug the coaxial cable coming from the VCR into the balun.

For photos of audio/video inputs, a threaded F-connector, and a balun, refer to the earlier section, "Hookup Using an Antenna."

For a simple setup, connect the cable from the wall to the cable box's input. Then connect the cable box's output to the VCR's input labeled "in from antenna." Set the VCR to channel 3 or 4 (whichever is set on the cable box). This will allow the user to watch and record any cable show. One disadvantage of this setup is that the user must remember to set the cable box to the correct channel in order to record a show while away. Also, this setup does not allow recording one show while watching another.

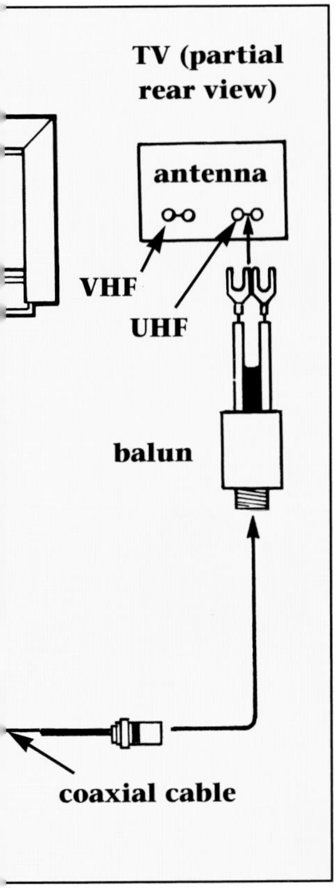

To get around this second problem, connect the cable from the wall to a cable splitter. This is a small box with three threaded F-connectors, available from most electronics stores. Now connect a coaxial cable between one of the remaining F-connectors on the cable splitter and the cable box's input. Connect another coaxial cable from the other F-connector on the cable splitter to the "A" input on an A/B switch (which is available from most electronics stores). Connect the cable box's output to the VCR's input labeled "in from antenna" and the VCR's output to the "B" input on the A/B switch. Finally, connect the A/B switch to the threaded connector on the back of the TV labeled either "antenna" or "ant/cable." Set the VCR to channel 3 or 4, whichever is selected on the cable box.

With this setup, the VCR can record any program selected on the cable box. The viewer can watch programs coming in from the cable box or tapes being played on the VCR by selecting "A" on the A/B switch and selecting channel 3 or 4, whichever is appropriate, on the TV. If desired, the VCR can be set to record whatever the cable box tunes in; then setting the A/B switch to "A" allows another channel to be tuned in and watched.

However, if all the channels on the cable system are scrambled, this scheme won't work—watching one cable channel while taping another simply can't be done unless two cable boxes are used. This connection, however, is simple. Just run the cable from the wall to the splitter. Connect one coaxial cable from the splitter to the input jack of the first cable box and connect this cable box's output to the "A" switch on the A/B switch. Connect another coaxial cable from the splitter to the input jack of the second cable box and connect this cable box's output to the VCR input labeled "in from antenna." Now connect a cable from the VCR output labeled "out to TV" to the "B" switch of the A/B switch. Finally, connect the A/B switch to the TV.

Using Audio/Video and S-Video Connections

If the TV has baseband audio/video or S-video connectors, use them—the picture will be much better. Use the appropriate set of instructions above to as-

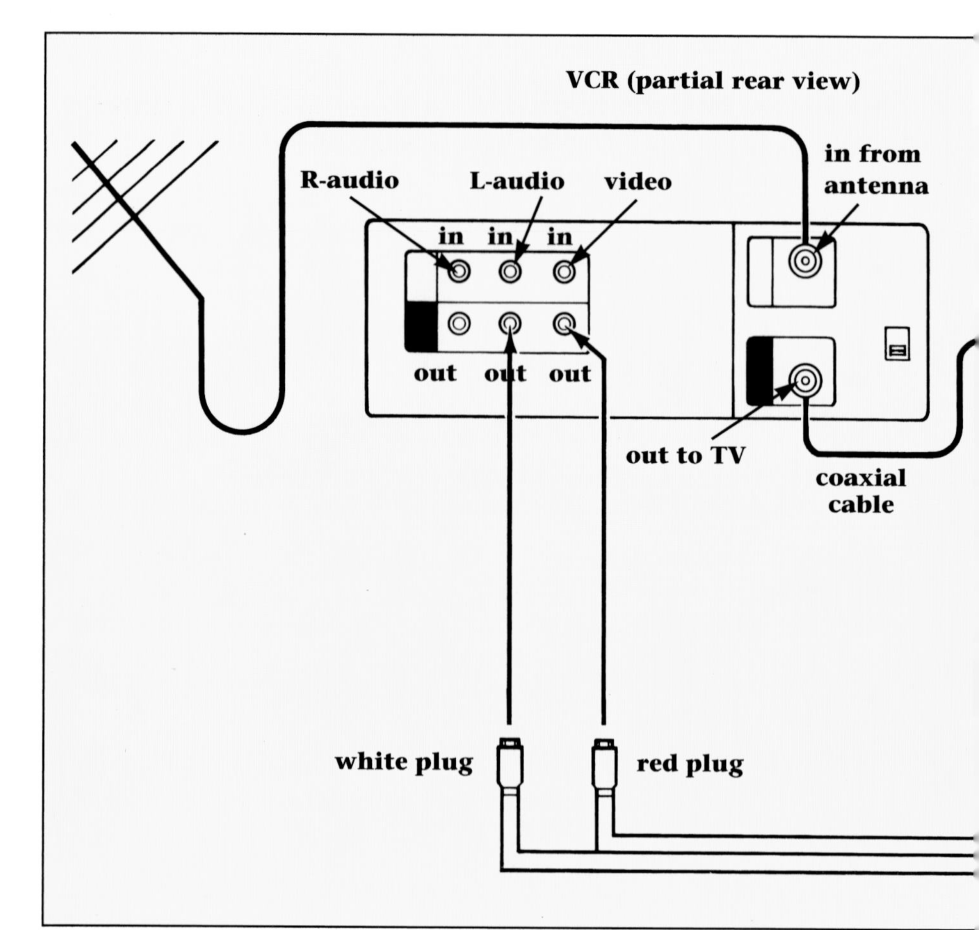

This schematic illustration shows how to hook up a monophonic TV to a stereophonic VCR.

semble the basic system. Then add the extra cables.

Monophonic TVs use one baseband input for audio and one for video. These inputs take the form of RCA-type jacks, which look like small cylinders about ¼ inch in diameter and ¼ inch long. The video jack will probably be colored yellow, and the audio jack will probably be colored white. The VCR should have matching jacks. Use a standard audio interconnect cable (with a red plug and a white plug at each end) to connect the VCR and TV. The red plugs are for video, and the white plugs are for audio. To watch a tape, select the TV's "auxiliary" input, which may also be labeled "line" or "video."

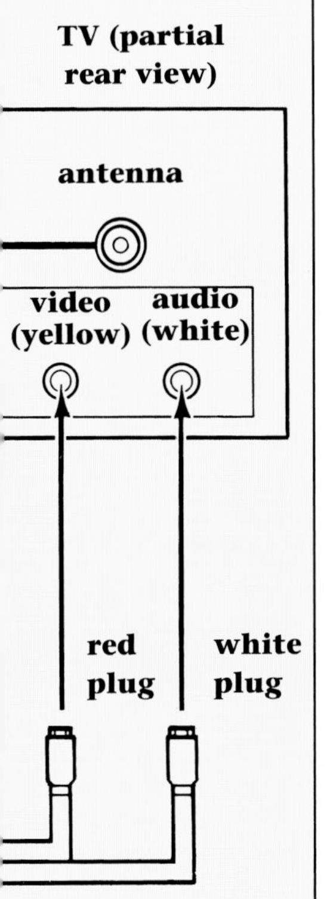

Connecting a stereo VCR to a stereo TV is almost as easy. Usually, the proper set of cables to use have three plugs at each end, one yellow (for video), one red (for right audio), and one white (for left audio). Just match and connect the plugs to the color-coded jacks.

To connect a stereo VCR to a monophonic TV, hook up the video cable. Then find the audio output on the VCR labeled "mono" (it's usually the white jack) and connect it to the TV's audio input. To connect a monophonic VCR to a stereo TV, connect the video cable. Then run a cable from the VCR's audio output to the TV's mono audio input (again, it's usually the white jack).

S-video outputs are found on S-VHS, Hi8, and ED Beta VCRs. S-video outputs carry the brightness and color parts of the video signal separately, which produces a slightly better picture. If the TV has an S-video input, connect the S-video cable from the VCR's S-video output to the TV's S-video

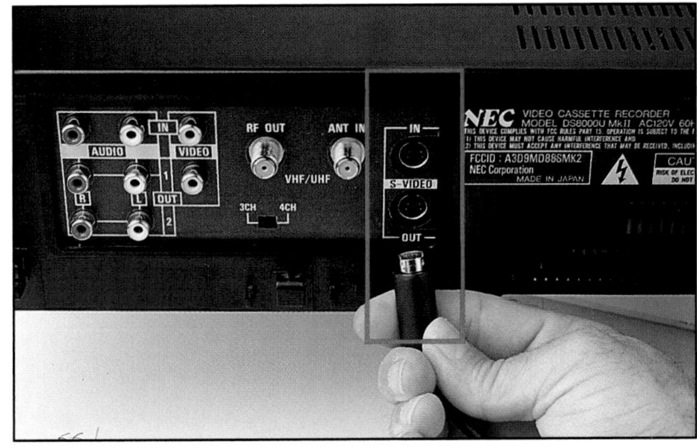

S-video jacks (box) carry the brightness and color parts of a video signal separately, producing a better picture.

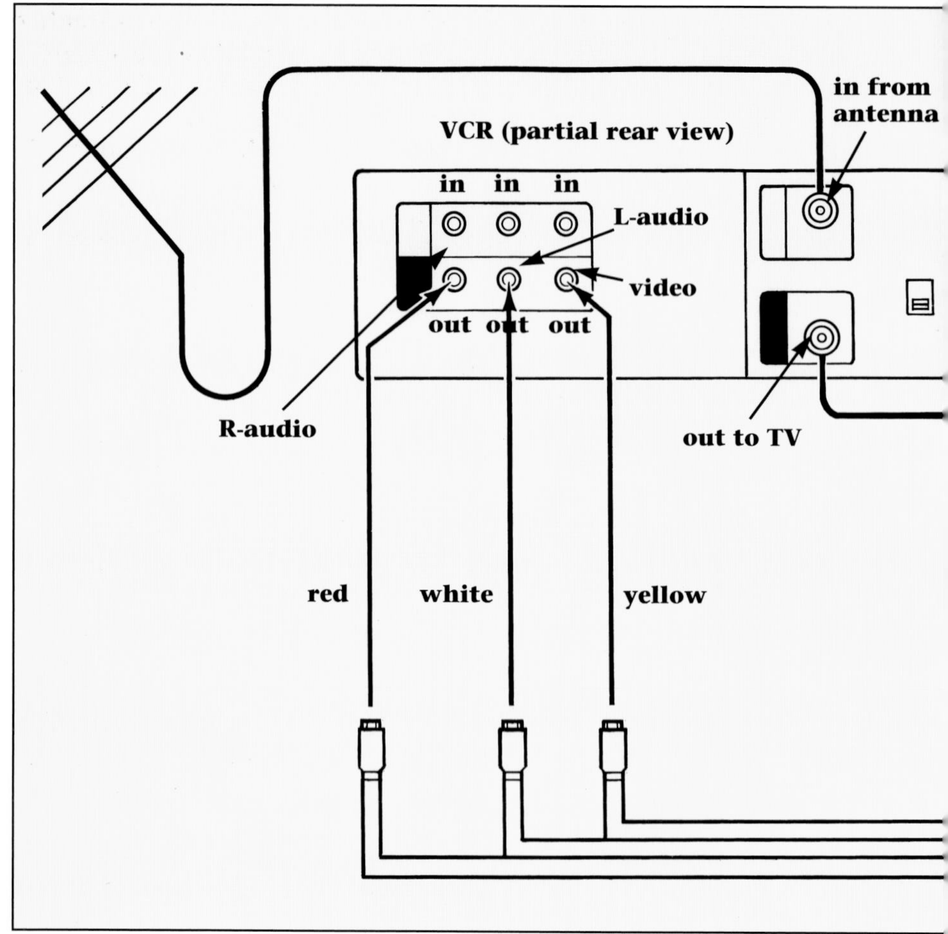

This schematic illustration shows how to hook up a stereophonic TV to a stereophonic VCR.

input. Connect the audio cables as described above. If the TV has no S-video input, ignore the S-video output on the VCR and use the standard video output.

Setting the Clock

The constant blinking "12:00" on the front of most people's VCRs has become synonymous with the problems of high technology. Getting that display to show what it's supposed to show—the correct time—has proved too taxing for many consumers' patience. Since many VCRs lose their time settings when the power goes out, the clock can't be set just once. In fact, the clock must be reset at least twice a year to change to daylight saving time and back. But setting a VCR clock is actually no more difficult than setting the time on a digital watch.

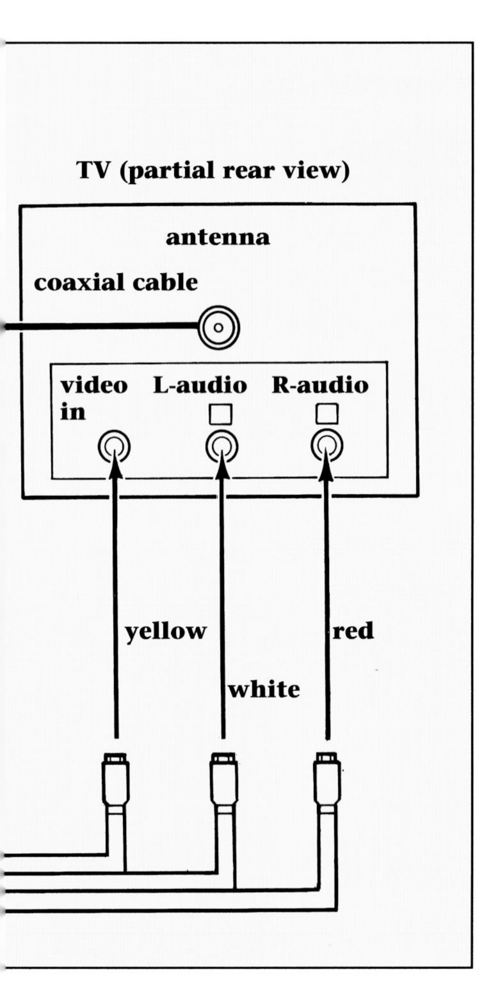

One element of the display—either the hour, the minute, the day, or the date—should begin flashing. If it is set correctly, push SELECT again to go to the next element. If it is not set correctly, push SET until the correct number comes up. Repeat the procedure until the VCR is set to the correct time and date. The button names may vary slightly from one VCR to another, but the procedure is almost certainly the same.

Newer VCRs use an on-screen menu to set the clock. This system is usually much easier to set than older systems. In fact, many VCRs automatically call up the clock set menu if the VCR loses power. And the remote control usually has a MENU button that accesses the clock set function. Once the menu is on screen, there shouldn't be any problems. Most menus ask for each piece of information and tell the user which controls to push to change the settings and to go to the next menu. Above all, don't be intimidated—most people who

Most older VCRs use front-panel buttons to set the clock. Usually, there are two buttons: One called SELECT and one called SET. To set the clock, hit the SELECT button.

Older types of clock-setting mechanisms have buttons, which most people find difficult to use.

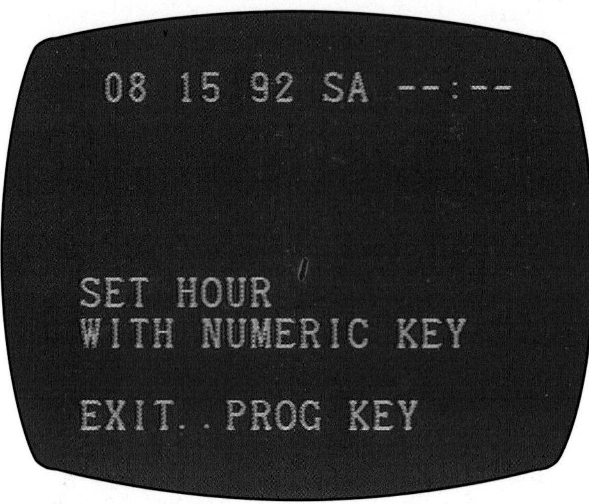

Newer types of clock-setting mechanisms have an on-screen menu that is much easier to set.

thought they could never set a VCR clock find they have no problem operating a menu system.

Operating Your VCR

Once the VCR is hooked up, take a good look at its controls (both on the front panel and on the remote control). The owner's manual should have a complete list of controls and their functions. But the easiest and most fun way to learn a VCR's operation is to put in a blank tape and play around.

All VCRs have a power (on/off) switch and a basic set of controls on the front panel, including FAST-FORWARD, REWIND, RECORD, EJECT, STOP, PLAY, and PAUSE. The FAST-FORWARD and REWIND controls double as scan buttons. Push one during play, and a forward or reverse speeded-up picture appears. There will also be channel selectors and a button labeled "VCR/TV." In the "VCR" position, the output will be from the VCR's tuner. In the "TV" position, the TV's tuner will function normally.

The number of additional buttons depends on the VCR. Some have many front-panel buttons, while others have only the ones named above and a few more. Sometimes a panel hides buttons that aren't used often.

Other buttons typically found on the front panel include SPEED SELECTOR, which activates one of two or three recording speeds; buttons labeled MENU, SET, and SELECT to control the menu system; INPUT SELECT, which sets the record input to the tuner or to the audio/video jacks in the back; INDEX, which electronically marks reference points on the tape; and TIMER ON/OFF.

Most hi-fi VCRs have an AUDIO MONITOR button, which selects either the hi-fi sound or the monophonic linear soundtrack. Some hi-fi VCRs have recording level controls, which adjust the volume of the sound being recorded.

The front panel should also have a button labeled OTR, ITR, OSR, or something similar. This button activates the one-touch record function. One-touch record starts recording on any channel being watched. Each touch of the button extends the recording time by one-half hour. There will also be a button nearby called STANDBY. This button delays the start of the one-touch recording by one-half hour with each press of the button.

Typically, the remote control will have many more buttons than the VCR. On most VCRs made since the mid-1980s, these buttons include all of those listed for the front panel, plus a ten-digit keypad for direct entry of channels.

If there is a menu system, the remote control will definitely have a set of controls for it. The remote

Although the front panels of most VCRs are similar, some differences do exist. A normal VCR looks much like the one shown at the top. A hi-fi VCR (bottom) is more complicated and has many more control buttons on the front panel.

may also have controls for FRAME ADVANCE, which moves the video ahead by one frame with each touch of the button; SLOW MOTION, which can usually be adjusted to one of several speeds; SPEED PLAY, which doubles the speed of playback; and COUNTER RESET, which returns the counter to zero. Many functions may be hidden in a menu system. These functions typically include programming controls, input selection, cable or antenna tuning band, audio monitor, and index search.

A typical on-screen menu provides a range of functions, usually more than can fit on the front panel of a VCR.

Playing a Tape

Many people use their VCR only to play rented movies. Playing tapes is so easy—almost like using an audio cassette player—that anyone who is at all familiar with audio or video gear should not even have to think about it. Just put in a tape and hit the PLAY button. The VCR automatically switches its output from the tuner to the tape, and the picture appears on the TV.

All newer VHS VCRs even do this automatically on prerecorded tapes. This is because prerecorded VHS tapes have no record tabs. The record

Most remote controls can make programming much easier for the viewer by accessing the on-screen menu.

tab is a piece of plastic at the rear of the cassette. If the tab is there, the VCR knows it's OK to record on the cassette. If no tab is present, the VCR won't record on the cassette. Instead, it will automatically play the tape when you insert it. The tab can be removed by prying it out with a finger or a sharp object.

The STOP button interrupts any operation. The PAUSE button puts the VCR in still frame mode. FAST-FORWARD moves the tape forward at a high speed, and REWIND moves it backward at a high speed, but both shut off the picture. Hitting either REWIND or FAST-FORWARD while the VCR is in the play mode causes the VCR to go into scan. The tape won't move as fast as in fast-forward or rewind, but a picture will appear. This is the mode to use when trying to find a certain spot on a tape.

All VCRs stop automatically when a tape is finished playing. Some of the latest decks automatically rewind tapes when they're finished playing. Some will even eject the tape and also shut themselves off.

A VCR uses an indicator to show it is recording (top, box). The VCR will not record unless the record tab—such as the one shown below on a VHS cassette—is intact.

Recording a Tape

The trick to recording a tape is to get it right before pushing the RECORD button. First, the tape that will be recorded on must have its record tab: On a VHS cassette, it's a tiny plastic tab about ¼-inch square at the rear. If the record tab is missing, the VCR won't go into record mode. A record tab on a VHS cassette can't be replaced, but a piece of tape over the hole will accomplish the same result. A sliding tab functions as the record tab on 8mm cassettes: The tab is red when the tape won't record and black when recording is OK. Most Beta cassettes have a record tab on the bottom near the hinge for the tape door, but a few have a sliding tab in the same place.

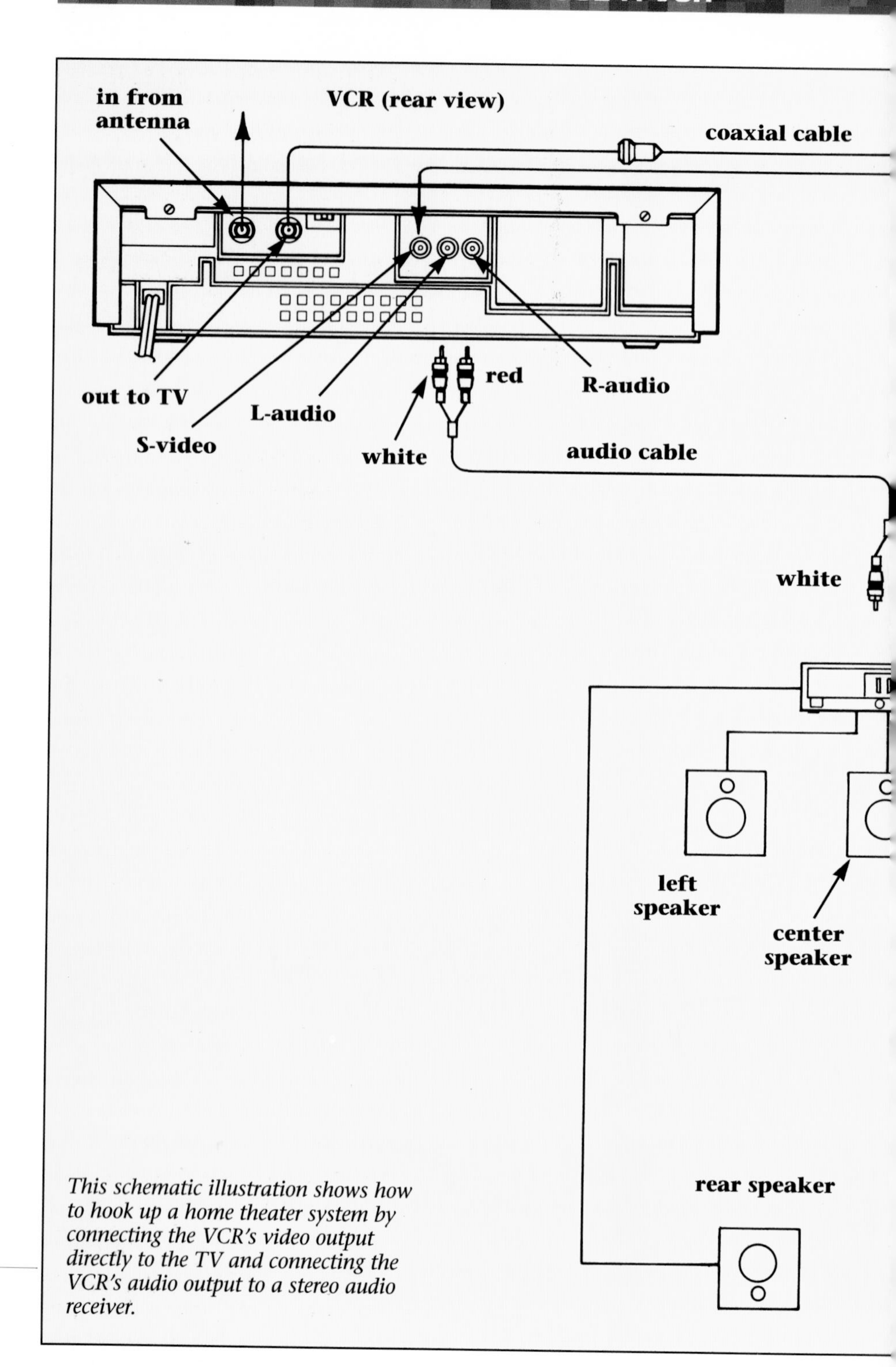

This schematic illustration shows how to hook up a home theater system by connecting the VCR's video output directly to the TV and connecting the VCR's audio output to a stereo audio receiver.

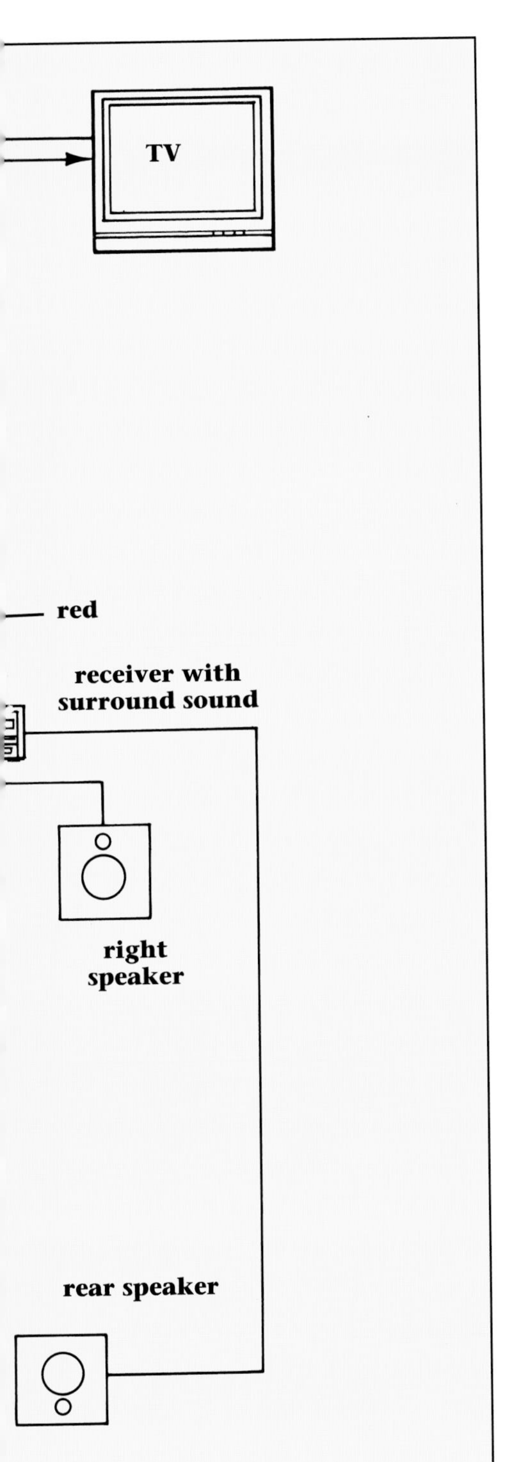

Next, select the desired record speed. For the best quality in VHS, use the standard (SP) speed. This provides two hours of recording time on the T-120 tapes found at most stores. To conserve tape, choose the four-hour long-play (LP) speed or the six-hour extended-play (EP) (also called SLP) speed. Some VCRs have a SPEED SELECTOR button; others hide the selector in an on-screen menu. Only standard (SP) and long-play (LP) speeds are available on 8mm VCRs. Later-model Beta VCRs have BII (two-hour) and BIII (three-hour) speeds. Some Beta decks, especially older models, have the BI (one-hour) speed.

If the TV is showing what is to be recorded, just push the RECORD button, and the VCR will record it. The VCR will record the selected channel or, if the line input has been selected, it will record from whatever is plugged into its audio/video input. If the VCR has a simulcast setting, it will record pictures from the TV tuner and sound from the audio inputs. On some VCRs, the RECORD and PLAY buttons may have to be pushed at the same time. A record light indicator will appear on the front of the deck. The PAUSE control can stop the recording momentarily, as when dubbing a tape and omitting the commercials. When recording is finished, use the STOP button. The REWIND and FAST-FORWARD buttons won't work while recording. When the VCR reaches the end of the tape, it will stop automatically. Some VCRs automatically rewind the tape;

others even eject the tape and shut off the power automatically.

Using a VCR in a Home Theater

One of today's most popular trends is the marriage of audio and video systems in a home theater. A big-screen TV provides a movielike picture; a high-quality surround sound system gives realistic sound. The centerpiece of many home theater systems is a hi-fi VCR.

The stereo hi-fi soundtracks on many tapes are encoded with Dolby Stereo, a special audio format that squeezes four channels of sound into two audio tracks. Audio/video receivers or surround sound decoders equipped with Dolby Pro-Logic decoding can extract the extra channels and send them to center and rear speakers.

To assemble one of these systems, most people use an audio/video receiver. These receivers have a surround decoder, amplifiers, and an audio/video switcher built in. Hook up the VCR and any other audio and video sources, connect five speakers, and it becomes a home theater.

Instead of hooking up the VCR to the TV via the audio/video cables, the VCR is hooked up to the receiver, and the receiver's output is connected to the TV. (To use one of these receivers as a video switcher, your TV must have audio/video inputs.) It is also possible to connect the VCR's video or radio-frequency (RF) output straight to the TV and connect the deck's stereo audio outputs to the receiver. With the RF connector, turn the TV's volume all the way down (or hit the MUTE button) when using the receiver. When using a separate surround sound decoder, hook the VCR's audio outputs to the decoder's inputs and connect the video output straight to the TV.

Even without a surround sound setup, much better sound can be accomplished by connecting the hi-fi VCR to the stereo system. Connect the audio outputs from the VCR to the auxiliary or video input of the stereo receiver. To play the VCR's sound through the stereo, turn the TV sound down and adjust the volume using the control on the stereo system. Be sure to set the TV between the speakers but don't put the speakers too close to the TV—they may interfere with the picture.

Some TVs have a variable audio output that allows the stereo system's volume to be controlled via the TV remote control. If the TV has this feature, the VCR can be connected to the TV in the nor-

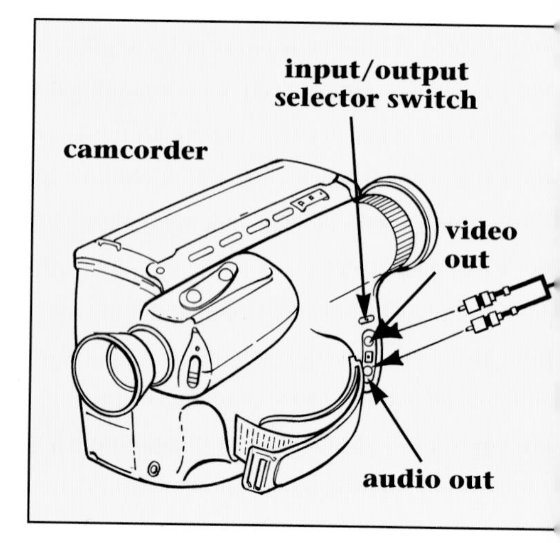

mal fashion with an audio/video cable. Then hook up a set of stereo audio cables between the TV's variable audio output and the auxiliary or video input on the receiver.

Making Copies of Other Tapes

VCRs are handy tools when used with camcorders and other VCRs. It becomes easy to make copies of a friend's tapes or personal camcorder tapes. Also, long, boring camcorder tapes can be edited into shorter, more entertaining tapes.

To make copies of another tape, connect the audio/video outputs of the camcorder or the second VCR to the VCR's audio/video inputs. Select the line or auxiliary input on the VCR. Put a tape in the VCR, hit the RECORD button, and then immediately hit the PAUSE button. Now begin playing the tape in the camcorder or the second VCR. Hit the PAUSE button on the first VCR to start recording.

To make an exact copy of the original tape, let both tapes run to the end. If only part of the original tape is to be recorded, hit PAUSE or STOP on the recording VCR to stop recording.

As your skill in copying tapes increases, you may want to try editing tapes. First, scan the original tape and write down the counter number at the beginning and end of each desired scene. Press the RECORD button on the VCR and then immediately hit the PAUSE button. Now start playing the tape to be recorded. When the counter number of the first desired scene comes up, hit the PAUSE button again to begin recording. When the end of the scene comes up, hit the PAUSE button again to stop recording. Repeat the process when the next scene comes up. Scenes can even be rearranged by fast-forwarding or rewinding while the recording VCR is in pause.

Video dealers also offer a variety of accessories to make even more creative editing possible. Generally, these accessories take the form of special effects devices that connect between the recording VCR and the camcorder or playback VCR.

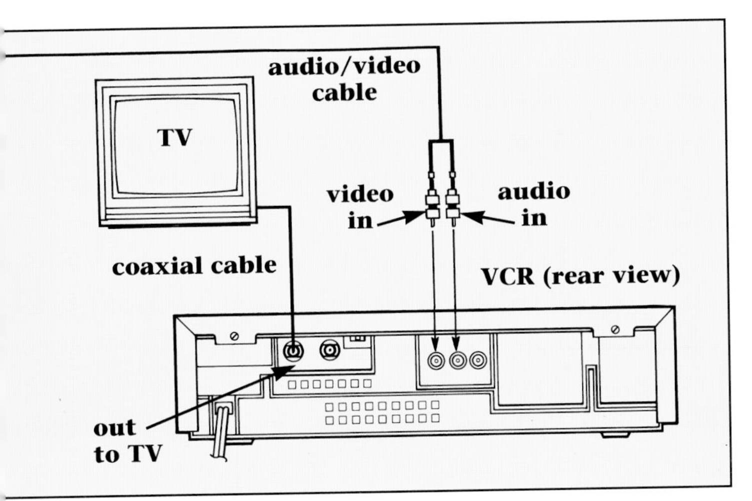

This schematic illustration shows how to hook up a camcorder to a VCR and TV to either record or view the camcorder's tape.

HOW A VCR WORKS

How Videotape Is Recorded and Played

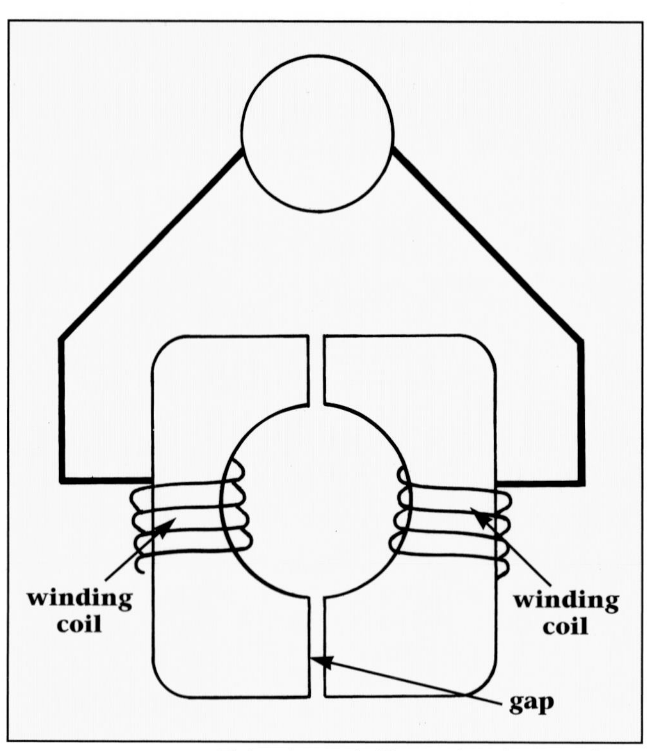

This schematic illustration shows a close-up of a video head. A video head is about half the size of a dime.

VCRs operate on the same basic principle as all tape recorders. They record electronic signals on a plastic tape coated with a magnetic substance. Usually, this substance is a metallic oxide (otherwise known as rust).

A video signal is basically fluctuating electrical energy. A TV set converts these fluctuations into a picture. VCRs record video signals by *boosting* (intensifying or making stronger) the signals and sending them to video heads, which have coils of wire wrapped around small pieces of metal. VCRs use two heads to record. When one head has finished its pass over the tape, the VCR automatically switches to the other one. The heads convert the fluctuating voltage into a fluctuating magnetic field, which appears in the gap between the small pieces of metal. As the head rubs against the tape, the magnetic field rearranges the magnetic particles on the tape into a

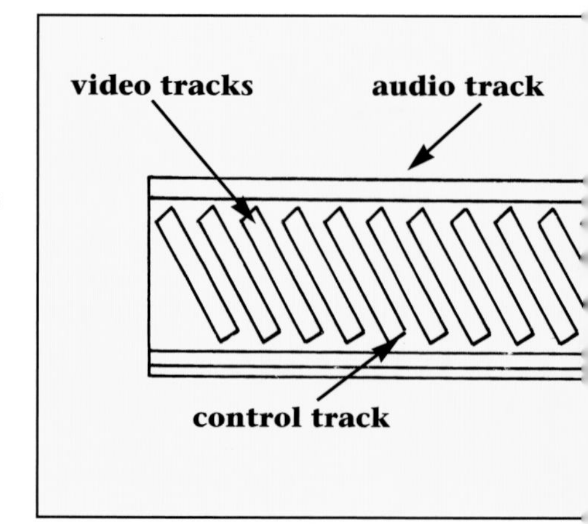

pattern that corresponds exactly to the fluctuations of the video signal.

To play a tape, the VCR drags the recorded tape back over the heads. This recreates the fluctuating magnetic field that was generated during the original recording. As the magnetic field passes through the head's coils, it is converted back into a video signal. This signal can then be sent to a TV for display.

The biggest difference between a VCR and an audio cassette deck is that the VCR's heads are mounted on a spinning drum. This drum rotates at an angle to the tape. The result is that the heads pass diagonally along the tape, creating a series of parallel tracks, each about 4 inches long. This arrangement is called helical-scan recording and is necessary because video signals cover a much wider range of frequencies than audio signals. In any tape recording (video or audio), the width of the space (the space from one edge of the tape to the other edge) used for recording doesn't matter that much, but the length (the amount of tape that passes over the head per second) of the recording space does. VHS helical-scan records 60 4-inch-long tracks per second, a total of about 20 feet of tracks per second, but only about 1³⁄₁₀ inches of tape moves across the head.

A separate track on the bottom of the tape—called the control track—carries pulses that inform the VCR exactly how much tape is passing by the heads. So even if the tape stretches, the deck pulls the tape along at exactly the right speed. A separate head, called the control head, reads the control track. Hi8 and 8mm tapes don't have a separate control track; their control signals—called pilot tones—are recorded along with the video signals. By automatically adjusting itself to receive the pilot tones clearly, the VCR optimizes video playback.

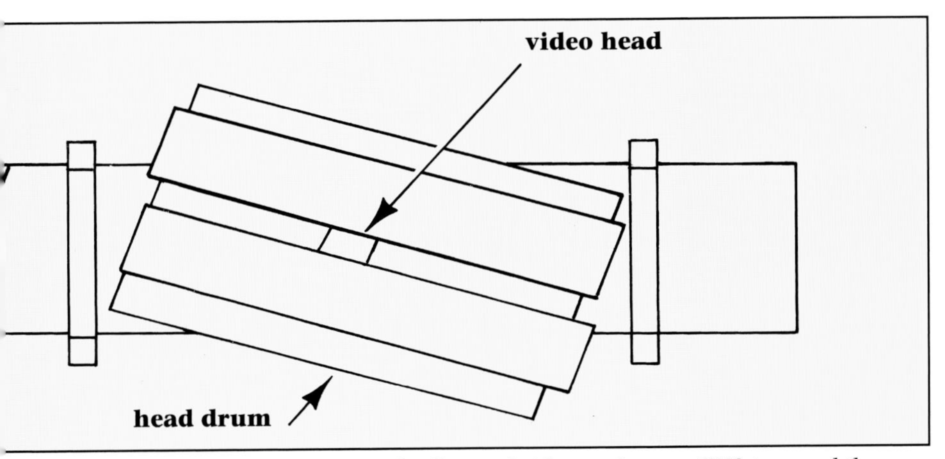

This schematic illustration shows the diagonal video tracks on a VHS tape and the tilted head drum that reads the tracks.

Important elements of the mechanical workings of a VCR: The tape loading mechanism (above) loads the cassette into the VCR. The transport (opposite page) wraps the tape around the video and audio heads.

On most VCRs, the control head also carries a linear audio head, which reads a monophonic audio track recorded along the top of the tape. Some VCRs have a separate linear audio head. Both the audio and control heads work like those on audio cassette recorders: They pass along the length of the tape, recording in a line. Some VHS VCRs made in the mid-1980s had stereo linear audio tracks, but the emergence of hi-fi audio for VCRs made these obsolete.

Hi-fi audio improves on the sound of the linear audio track by placing the sound with the video tracks. The hi-fi audio heads ride along with the video heads on the spinning head drum. VHS hi-fi audio is recorded just before the video, then the video is recorded on top of the audio track. The audio signal is not erased because it is made up of lower frequencies than the video signal. Since the video signals comprise higher frequencies than the audio signals, the video signals do not penetrate as far into the tape. Thus, the video signals do not appreciably affect the audio signals. Also, the audio heads are set at a different angle than the video heads to achieve better differentiation between the signals, thus maximizing the quality of the signal. Beta hi-fi decks record the audio signal along with the video signal, using the video heads.

Before a videotape is recorded, it passes by an erase head. This head covers the full width of the tape. The erase head uses a magnetic field to mix up the pattern of any previous recording. This prevents old pictures and sounds getting mixed with the new picture and sound. Some VCRs, including all 8mm and Hi8 models, have a flying erase head, which rides on the video drum.

How the Tape Moves Around Inside the VCR

The mechanism that moves the tape inside the VCR is called the transport. It performs several tasks. It pulls the tape out from the cassette into the VCR; it wraps the tape around the head drum and against the audio/control and erase heads; and it pulls the tape from one reel, through the maze of heads, onto the take-up reel.

Early VCRs used a top-loading design: The user inserted the cassette into a basket and then pressed the basket down, lowering the tape into the transport. When front-loading VCRs were introduced in 1980, top-loaders quickly faded from the scene. But many top-loaders are still running strong after a decade of use.

Front-loading VCRs use a tape basket, which is a frame made from sheet metal and plastic that hides behind the cassette hatch on

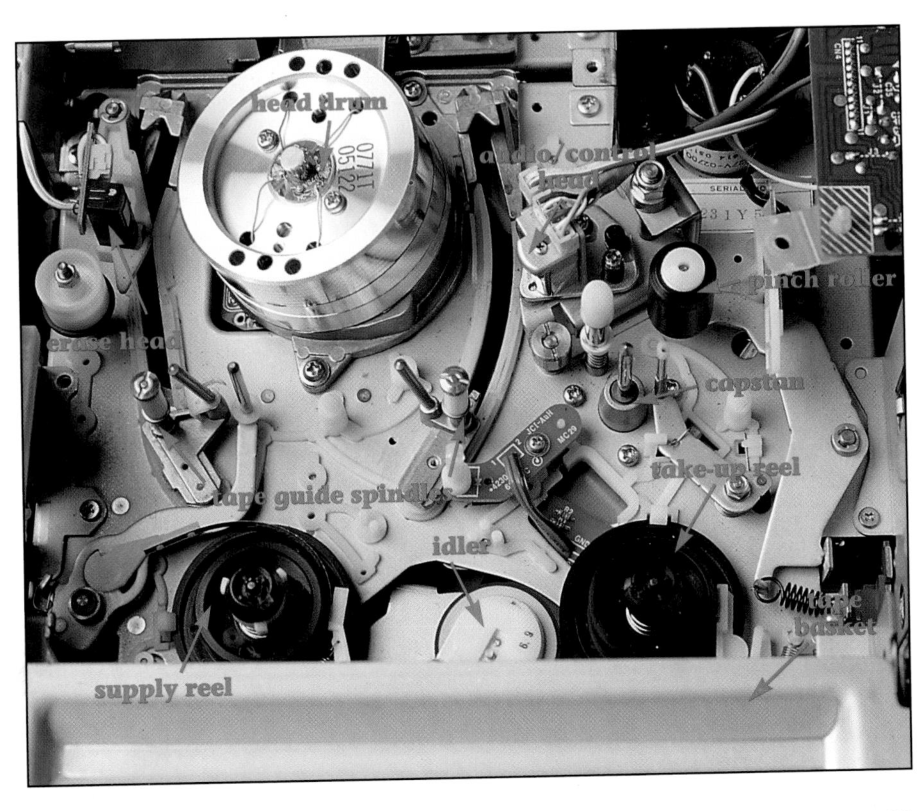

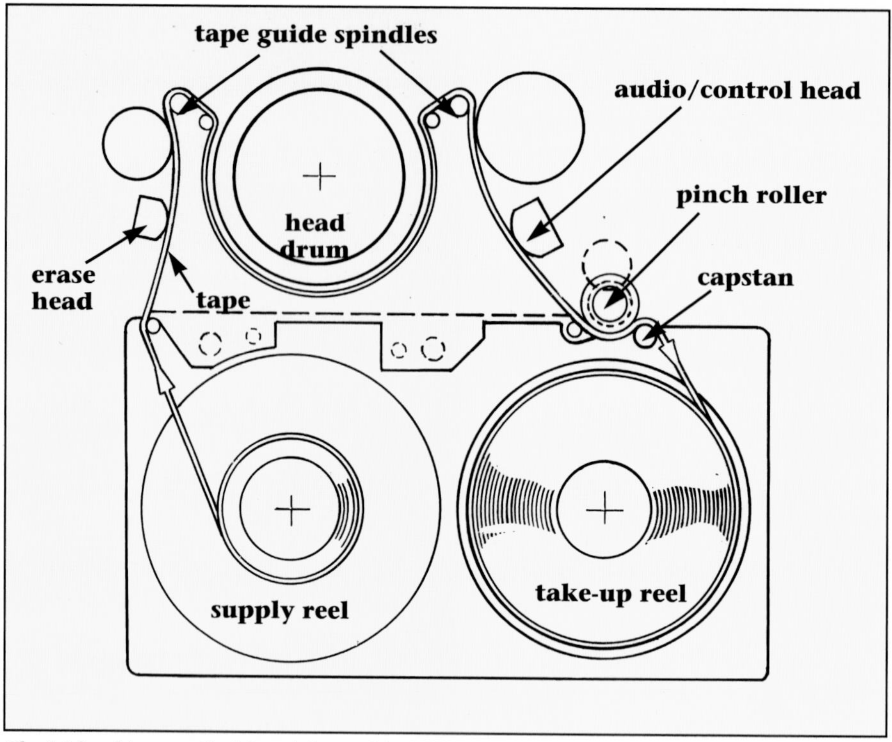

The M-load tape path (above) is used in VHS VCRs; the U-load (opposite) is found in Beta VCRs.

the front of the VCR. The basket houses a tiny switch that closes when a tape is inserted. When this switch closes, a motor inside the VCR pulls the cassette fully into the basket and then lowers the basket onto the transport. A pin alongside the basket releases the catch that holds the hatch shut on the front of the tape cassette.

As the tape is lowered, it comes into contact with several important transport parts. The spindles fit into the large splined holes on the reels inside the cassette. A motor turns these spindles, thus moving the tape back and forth. A pin pushes up through a hole in the bottom of the cassette to release the brake that locks the reels in place.

On VHS and 8mm VCRs, a sensor lamp also enters through the bottom of the cassette. This lamp helps the VCR know when the tape has reached the end, so it can stop the function it is performing and avoid damaging the tape. During operation, the tape blocks the light from the lamp. But clear plastic leaders attached to the ends of the tape let the light shine through when the tape reaches the end. The light reaches one of two tiny optical sensors positioned at each end of the cassette. When light hits these sensors, the VCR stops. Beta VCRs use a magnetic sensor and metallic strips on the tape to accomplish the same thing.

Two guide posts and a capstan enter behind the length of tape stretched across the front of the cassette. The posts pull the tape out of the cassette and wrap it around the head drum. A rubber pinch roller presses the tape against the capstan, which is connected to a motor. As the capstan turns, it pulls the tape through the VCR.

VCRs use different types of tape paths. Beta VCRs use a U-load transport, so called because one post moves in a circular pattern, pulling the tape past the head drum and then back around in a U-shape. VHS and 8mm VCRs use an M-load transport, which uses two posts to wrap the tape around the drum, thus forming an M-shape.

Beta VCRs keep the tape wrapped around the head drum all the time. Because of this, fast-forwarding and rewinding the tape can prematurely wear out both the tape and the heads. Most VHS VCRs avoid this wear by unwrapping the tape from the head in the fast-forward, rewind, and stop modes. The price for this is speed: The VCR doesn't operate as fast because unwrapping the tape takes a couple of extra seconds. Also, because the tape doesn't touch any of the heads in these modes, the VCR can't display accurate counter information.

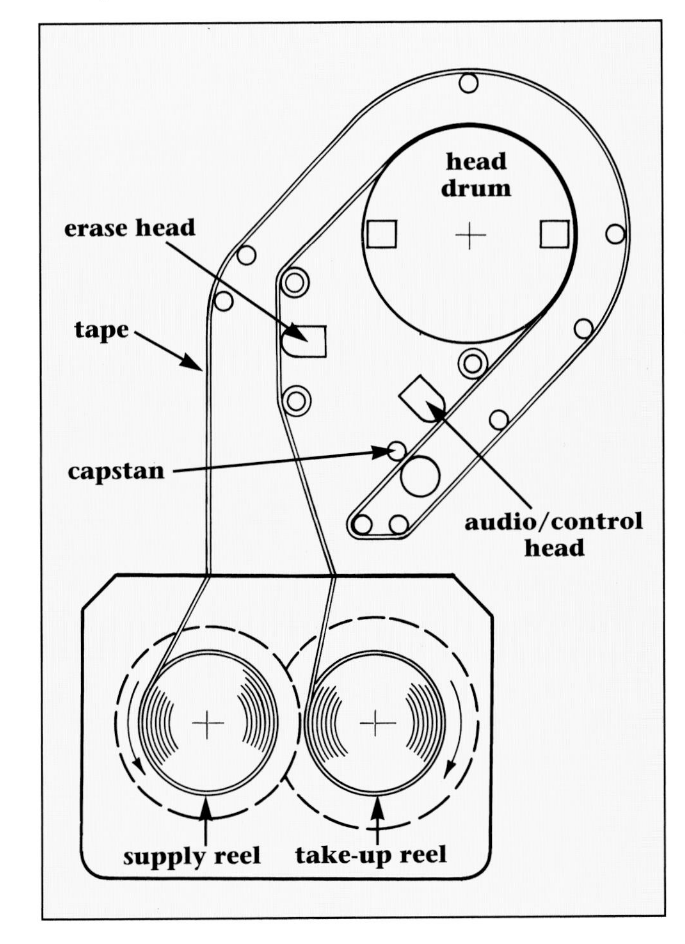

Most of today's better VHS VCRs use a half-load mechanism, which unwraps only a part of the tape in the fast-forward, rewind, and stop modes. Since the tape still comes in contact with the control head, the VCR can display a real-time counter that supplies the tape position in hours, minutes, and seconds. Half-load mechanisms also operate much faster than standard VHS transports.

Inputs and Outputs

For many people, one of the most confusing aspects of using a VCR is hooking it up in the first place. An understanding of how a VCR's inputs and outputs work will make hooking up any VCR—in any system—easy.

All VCRs have at least two types of inputs, one for baseband video signals and one for radio-frequency (RF) signals. Both carry the same information. The difference is that the RF signals are modulated, or shifted up to a much higher range of frequencies. RF signals combine the video and audio information and are the same as what you get from a cable or antenna. Thus, a cable or antenna will usually be connected to the RF input, which uses a jack called an F-connector.

While the RF input can carry many channels of video, the baseband input carries only one. The frequency of these signals must be demodulated, or shifted back down to the baseband range, before the signals can be recorded by a VCR or displayed by a TV.

The baseband video input carries video signals at their original frequencies. Use this input to hook up a second VCR, a camcorder, or a laser disc player to the VCR when making copies of a video. Next to the video input is a baseband audio input, which can accept audio signals from another VCR, a camcorder, or an audio system. Stereo VCRs have inputs for right- and left-channel audio. Some VCRs

F-connectors (box) can accept cables that carry radio-frequency (RF) signals, which carry many video channels.

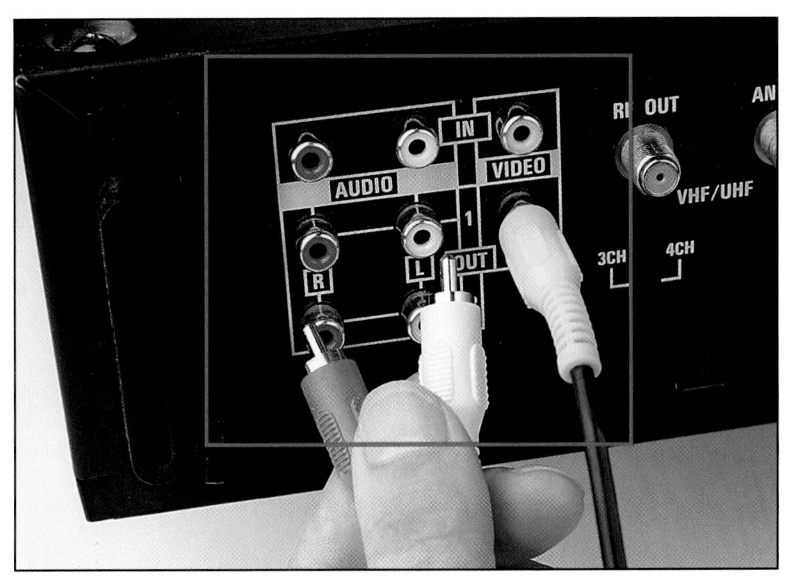

Baseband inputs, both video and audio, accept only one signal, but the signal is of a higher quality than RF signals.

have two sets of baseband inputs, one on the front panel and one on the back. All of these use RCA-type jacks, similar to those found on the back of a stereo system. Some VCRs also have a minijack for a microphone.

S-VHS, Hi8, and ED Beta VCRs have an S-video input in addition to RF and baseband inputs. An S-video input accepts baseband video that has been separated into brightness (or luminance) and color (or chrominance) signals. It is sometimes called a Y/C input. Using this input reduces *hanging dots*—the crawling effect seen at the bottom of horizontal lines on the TV screen.

The outputs of VCRs tend to match their inputs. The RF output sends signals from a tape or the VCR's tuner to the TV set. A selector switch sends the RF signal over either channel 3 or channel 4, whichever channel is not used by broadcasters in the local area. If the VCR is switched off, the signal coming in from the antenna or cable is simply passed to the TV, so the TV operates normally.

The baseband video and audio outputs connect either to TVs with baseband inputs or to another VCR for copying or editing. The signals coming from these outputs are of higher quality than the signals coming through the RF output, so use these outputs whenever possible. S-VHS, Hi8, and ED Beta VCRs also have an S-video output, which should be used if the TV accepts S-video signals.

Electronics

Although VCRs rely heavily on mechanical parts in recording, the electronic

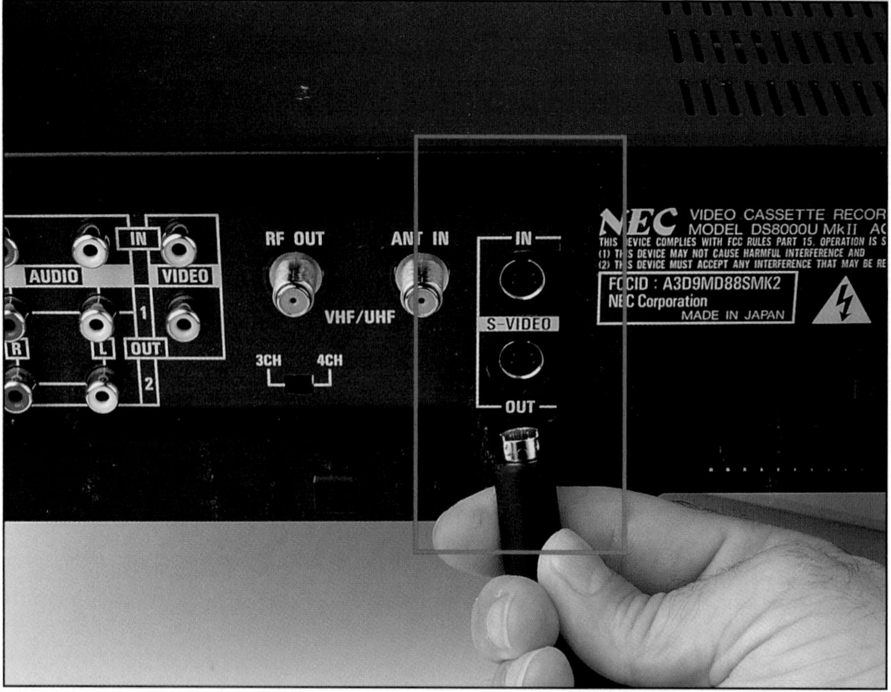

S-video inputs (box) are found on S-VHS, Hi8, and ED Beta VCRs. These inputs help eliminate some annoying video problems.

parts are equally important. Besides the recording electronics already discussed, VCRs also have a tuner, just like the one inside a TV. The tuner selects one of the many channels coming in from the antenna or cable through the RF input. The tuner then converts the channel's signal to baseband video and audio. Stereo VCRs have tuners with MTS (multichannel television sound) decoders, so they can receive and record stereo sound from television stations.

VCRs also carry a timer, an electronic clock that not only keeps time but also activates a variety of VCR functions. When the timer is set, it tells the VCR at the appropriate time to power up, tune in the correct channel, start recording, and then stop recording and power down after the show is over.

Today's VCRs have so many features the designers can't fit enough buttons on the front panel to control them all. Instead, onscreen control systems with a remote control are used. Almost all VCRs now have infrared remote controls. An infrared remote control runs a VCR by sending out precisely timed pulses of invisible infrared light. A window on the front of the VCR detects these pulses and interprets them as commands. Each VCR brand uses its own set of pulse codes, which is why a VCR remote will usually operate any VCR of the same brand.

THE CARE AND MAINTENANCE OF A VCR

The Right Environment

VCRs need to be kept in the proper environment to function properly. Fortunately, VCRs were designed to operate in the same place people usually operate: in the home. But the typical residence holds hazards for VCRs that can degrade their performance, either slowly or abruptly.

The most dangerous hazard is moisture. If moisture gets inside a VCR, it can short out the circuits, attract dust and debris, or cause transport parts to stick to the tape. These problems can result in poor pictures and sound, "eaten" tapes, and even complete breakdown of the VCR.

Most moisture contamination of VCRs occurs through spills. Obviously, keep drinks well away from the VCR and the rest of the audio/video system. Don't let children near the deck with cups and glasses. Also, avoid putting plants

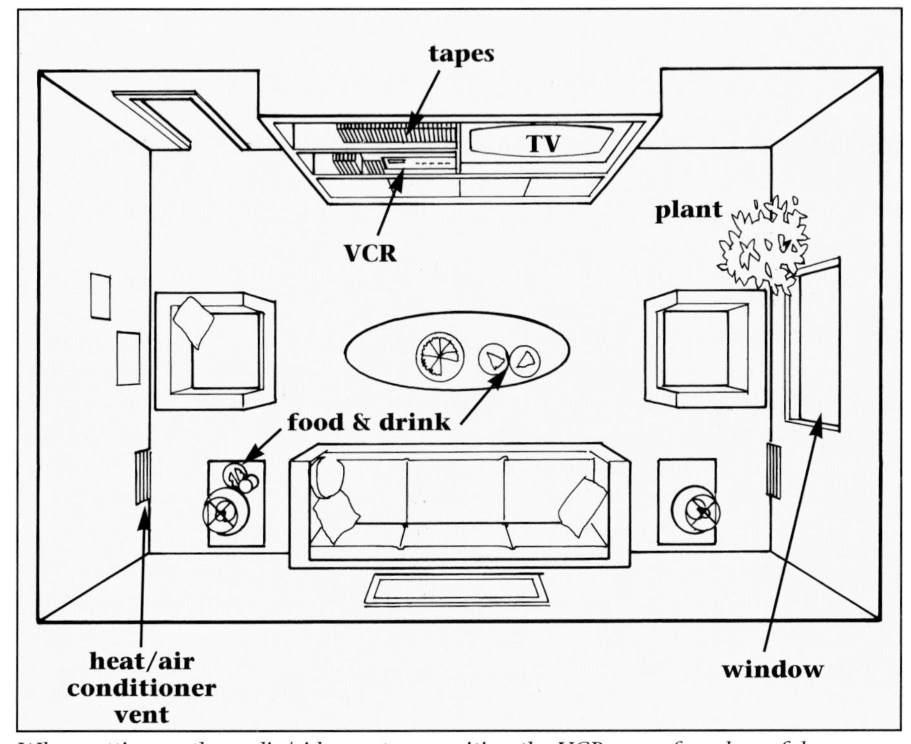

When setting up the audio/video system, position the VCR away from harmful elements: direct sunlight, heat and A/C vents, plants, and food and drink.

near the VCR—it's easy to spill the watering pail or overfill the pot. Also keep the VCR away from air conditioners or cool-air vents, because cool air can cause condensation to form on the deck in humid environments.

If someone accidentally spills liquid on the VCR, wipe it off immediately. If it appears the liquid found its way inside the VCR, remove the top (the procedure is described later in this chapter) and soak up the liquid with paper towels or a rag. Be careful not to touch the video heads.

The second most common contaminant is dust. But keeping dust out of the VCR is simply a matter of keeping the VCR away from dusty areas. Fortunately, few areas around the house are so dusty they're likely to cause problems for the VCR. As long as the surrounding area is kept relatively clean and dusted and the VCR is dusted occasionally, it's unlikely the VCR will be seriously contaminated with dust. In fact, it's far more likely that dust particles will be carried into the VCR via videotapes.

Also avoid placing the VCR in an area where it will encounter direct sunlight or abnormal heat. Both can warp the plastic parts of the VCR. The inside of the VCR can turn into an oven in direct sunlight. Also place the VCR well away from heating vents. Be sure to leave a couple of inches of air space around it so it can cool. If the VCR has cooling vents on the top, sides, or back, don't obstruct them.

If the VCR gets too hot, the picture quality goes down and the motors and circuitry may not last as long. Sunlight can fade the finish of the VCR's cabinet, which won't affect performance but will destroy its appearance. An environment that is too hot is especially bad for the VCR if the user doesn't always eject a tape when finished viewing it or if the user keeps a blank tape in the deck at all times for unexpected recordings. Heat can damage and often completely ruin a tape.

Maintaining a VCR

To most people, "maintenance" probably means the constant effort it takes to keep a car running smoothly—checking tire pressure and fluid levels, washing it, tuning it up every year, and dozens of other tedious chores. But maintaining a VCR is much simpler: All it really needs is an occasional cleaning, which might take 30 minutes a year.

Cleaning is more important with a VCR than with audio tape recorders for two reasons. One, since the video heads are much smaller than most audio heads, they are more affected by dirt. Two, the spinning heads of a VCR exert more wear and tear on a video tape than the stationary heads used in audio recording. This wear causes the tape to leave bits of itself on the video heads.

If the video heads in your VCR are dirty, the effects can usually be seen right away. The dirt may show up as white streaks running across the picture, or as a very snowy picture, with white static covering the image from the tape. Usually, the sound quality on the tape will be fine—only the picture will be bad.

Because video heads are more delicate than audio heads, don't go overboard on cleaning. In fact, it's usually OK to wait until a problem develops to clean the VCR.

Some of the latest VCRs have automatic head cleaners, which are tiny rollers that briefly rub against the video heads every time a tape is inserted or ejected. There shouldn't be any worry about cleaning the heads of these VCRs.

The other parts of the transport can (and probably should) be cleaned more often. The more dirt removed from the transport, the less chance there is of dirt finding its way onto one of the heads. If cleaning is done correctly, these other parts won't be harmed. If the VCR starts "eating" tapes, the transport should definitely be cleaned; chances are the tape is getting hung up on a gummed-up part.

Head-Cleaning Tapes

There are two basic methods of cleaning VCR video heads. Either take it apart and clean it by hand or simply pop in one of

One way to clean video heads is with a dry-type cleaning tape.

When using a dry-type cleaning tape, the TV screen will show blurred text (top). When the cleaning tape has finished its job, the text will be readable (right).

the many available cleaning tapes. Both methods have advantages and disadvantages.

One main advantage of cleaning tapes is convenience. With most cleaning tapes, just insert one in the VCR and let it play until the heads are clean. But these tapes don't clean as well and are not as gentle on the VCR as a manual cleaning.

Wet-type cleaning tapes, which use cleaning fluid, are notorious for causing as many problems as they solve. The best cleaning tapes are the dry type. These do a good job of cleaning. Because of their abrasive nature, however, dry-type cleaning tapes can cause premature head wear if used too often.

To use a dry-type cleaning tape, just insert it into the VCR and watch the TV screen. If the heads are dirty, blurred text will appear in the middle of the screen. Once the heads are clean, the text will be readable and will tell the user to stop the VCR immediately. Don't run the tape any longer than necessary: It will just wear down the heads.

On occasion, the VCR may develop a *head clog*—a large, stubborn piece of debris stuck on one of the heads. One cleaning won't get it off. Sometimes, a head clog can be removed without taking the VCR apart: Simply run a cleaning tape through the deck three or four times. If the picture is still bad after a few runs with the head cleaning tape, clean the heads manually.

Sometimes, a VCR may develop a head clog, the effects of which can be seen above.

Manual Cleaning

Cleaning the VCR's heads by hand is better than cleaning tapes for two reasons. One, the heads are cleaned more thoroughly. Two, the rest of the inside of the VCR can be cleaned at the same time, removing dirt that could someday find its way onto one of the heads.

To clean the heads by hand, the user must get inside the VCR. That's not as frightening as it sounds. On many VCRs, a basic cleaning job only demands removing the VCR's top cover. For most front-loading VCRs, that involves removing only four or five screws. A piece of sheet metal or a small circuit board may also have to be removed.

If the VCR is still under warranty, take it to an authorized service center for cleaning or try a dry-type cleaning tape instead of cleaning it manually. This is because *opening the VCR usually voids its warranty.* If something goes wrong with the VCR after it has been cleaned but before the warranty is up, the user will have to pay the repair bill.

Materials

Before opening the VCR, collect the needed tools. The only conventional tools required are slotted and Phillips screwdrivers in var-

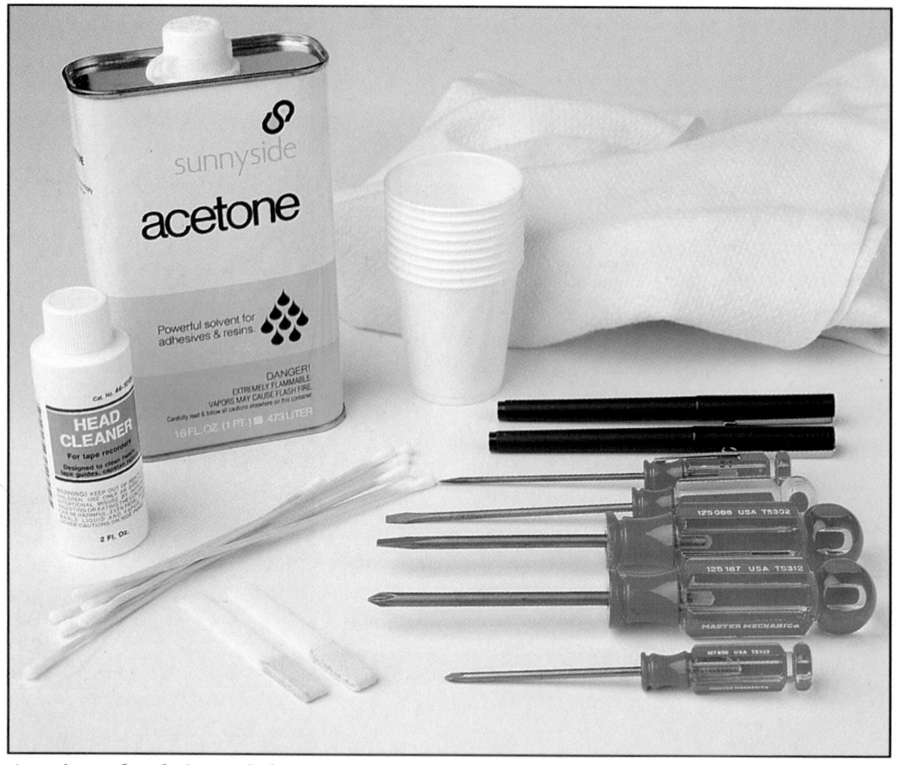

A variety of tools is needed to manually clean a VCR, but all are easily obtained at hardware or electronics stores.

ious sizes (to be prepared for any type of screw encountered inside the VCR). Be sure to use screwdrivers that fit right: The screws inside VCRs are usually very tight, and the screw heads shouldn't be damaged by using a screwdriver that is too small.

Head cleaning sticks designed especially for use on video heads will also be needed. These sticks have tips made of chamois and are available at electronics stores. Never, under any circumstances, use cotton-tipped swabs to clean video heads, not even the type of swabs designed for cleaning audio heads. These cotton swabs can snag on the heads and pull them out of alignment. But cotton-tipped swabs designed for audio tape head cleaning are useful for cleaning other parts of the VCR. A clean, lint-free rag is useful. Head-cleaning fluid, available at electronics stores, should be used to clean the heads. Alcohol or acetone will also work. Finally, grab several paper or plastic foam cups, a piece of paper, and a pen.

Taking the VCR Apart

First, pull the machine out of the system and unplug it. Before proceeding further, double-check to make sure it's unplugged: It's easy to forget, and the voltage inside a plugged-in VCR can be dangerous.

Don't reach for the screwdriver yet. Instead, grab that pen and paper and make a simple drawing of the VCR, marking the locations of the screws that hold the top of the VCR in place. These screws are usually on the side and back of the deck. Number the locations

Only a few screws on the side and back must be removed to get inside the VCR (right). Once the screws have been removed, the top of the VCR should be removed; the top may need to slide back a little before it can be lifted (bottom).

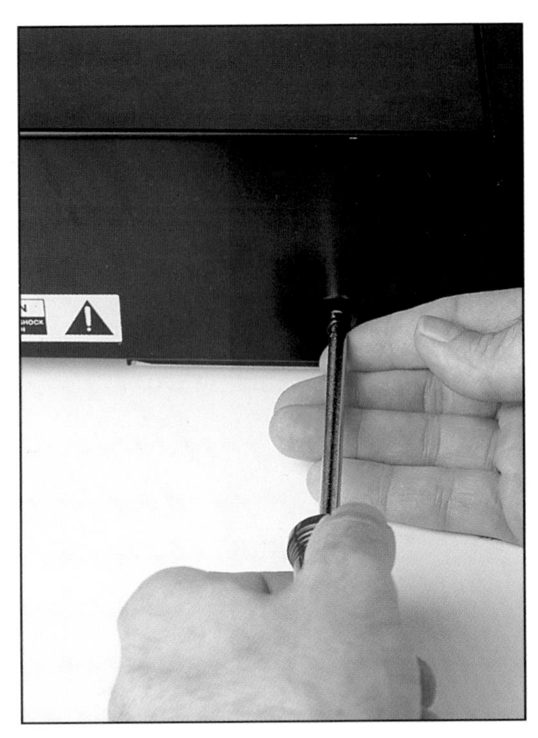

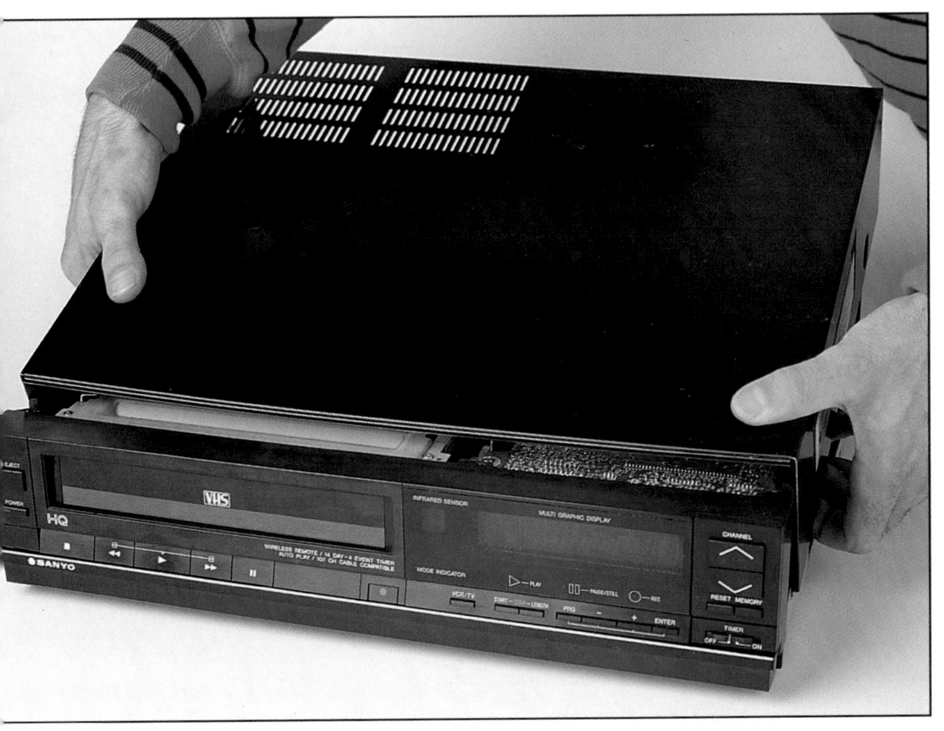

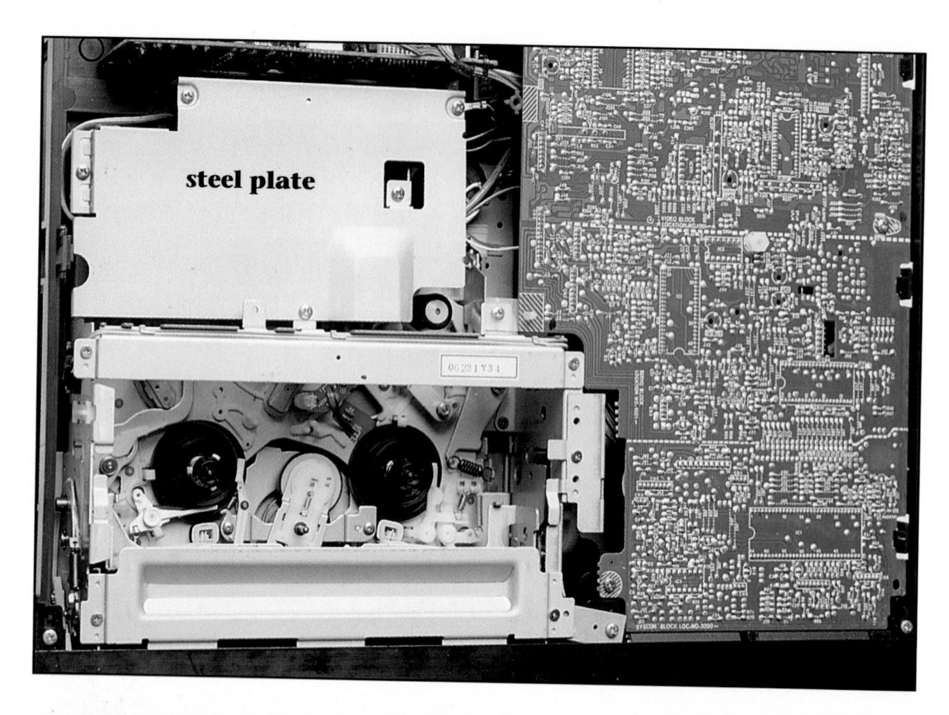

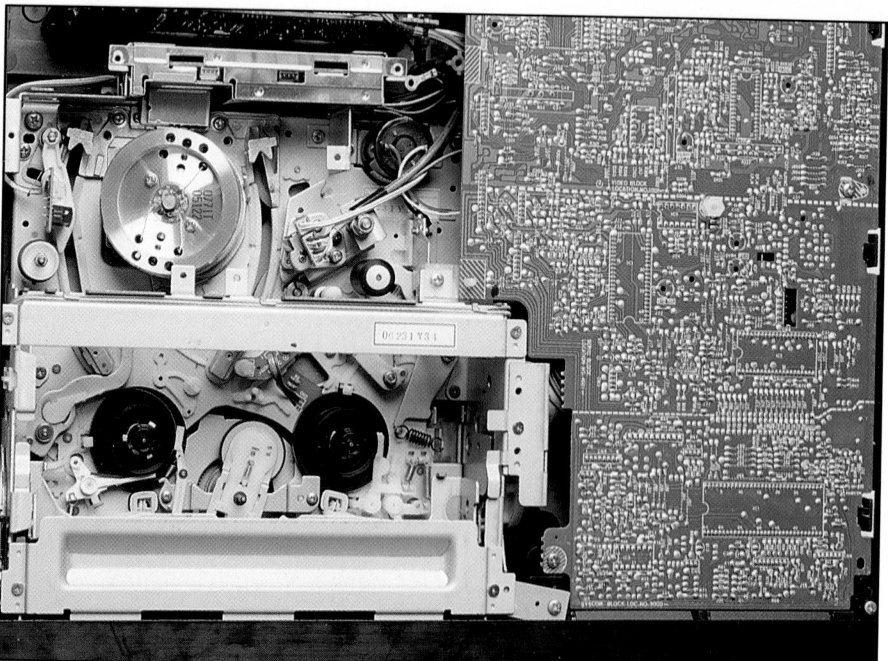

Once the top is lifted, a steel plate or circuit board may have to be removed as well (top). With the steel plate removed, the mechanical parts of the VCR will be exposed and ready for cleaning (bottom).

of the screws on the drawing. As each screw is removed, write its number on a cup and drop the screw into the cup. This may seem like a lot of trouble, but it will prevent stripping out the holes in the VCR chassis by trying to force the wrong screw into the wrong hole. Place the cups where they're unlikely to be knocked over. Be especially careful not to drop any of the screws into the VCR mechanism.

Once all the screws are out of the top, carefully remove the top. It may have to be slid slightly backward. Put the top aside and look at what is now uncovered. If all the transport parts can be seen, it's time to start cleaning.

The video heads (arrow) in a VHS VCR are extremely small and are located on the head drum.

On many VCRs, however, a circuit board or a steel plate may cover the transport. If there is a circuit board or steel plate, make a drawing similar to the one made for the top. Remove the screws and place them in their own cups. Note carefully how the board or plate fits. Circuit boards will have wires attached. Don't try to disconnect the wires—just gently lay the board aside, securing it temporarily with electrical tape.

VHS and 8mm VCRs

The video head drum is the large, shiny, tilted cylinder seen at one end of the transport mechanism. The heads, however, are harder to find. Note that the top half of the drum rotates, while the bottom part is fixed. Rotate the top part and watch along the slot between the two halves. Tiny, slotted pieces of metal along the bottom edge of the top half of the drum will appear. These are the video heads. There can be two, three, four, or even six of them. In addition, 8mm and VHS hi-fi VCRs will have two extra heads for audio. VHS VCRs for editing and 8mm VCRs will have

Use a head cleaning stick to clean the video heads; move the head drum, not the stick.

one or two erase heads mounted on the drum as well. All these heads must be cleaned.

To clean a head, wet the head cleaning stick with cleaning fluid. Place the stick against the side of the drum and press the tip of the stick against the groove running between the top and bottom halves of the drum. Now rotate the top part of the drum back and forth so that the head wipes against the pad of the cleaning stick. Move the head drum only: Don't move the stick or the head could be knocked out of alignment. Repeat this procedure for each head on the drum.

Beta VCRs

The head drum on a Beta VCR looks much like that of a VHS VCR: It's a large, shiny cylinder. Unlike the drum head on VHS VCRs, the drum head on a Beta deck doesn't rotate. Instead, an armature inside the drum rotates. This armature rides on an axle that sticks out from the top of the head drum.

Press a finger to the top center of the head drum and turn it. The axle will rotate beneath the finger. As the axle rotates, tiny slotted pieces of metal will move through the slot between the top and bottom halves of the head drum. These pieces of metal are the

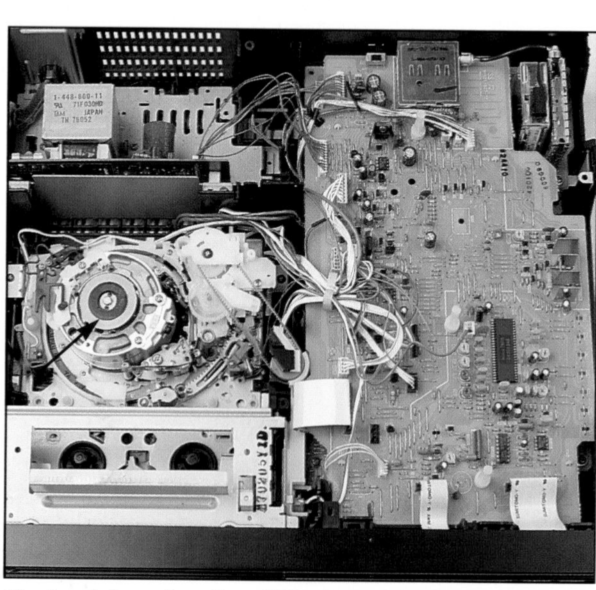

The head drum in a Beta VCR (arrow) is slightly different from a VHS head drum; note the axle in the middle of the head drum (arrow).

The video heads (arrow) in a Beta VCR (top) are also very small and are located on the head drum. Clean the video heads (bottom) of a Beta VCR in the same manner as cleaning the heads of a VHS VCR.

video heads. There may be two or four of them. Beta hi-fi VCRs will have two audio heads riding through the slot. Beta VCRs for editing will also have one or two erase heads mounted on the armature. All these heads need to be cleaned.

To clean a head, wet the head cleaning stick with cleaning fluid. Place the cleaning stick against the side of the drum and press its tip against the diagonal groove running between the top and bottom halves of the head drum. Now press a finger against the top of the axle and rotate it so that the head wipes against the pad of the cleaning stick. Move the axle only: Don't move the stick or the head could be knocked out of alignment. Repeat this procedure for each head on the armature.

Cleaning the Rest of the Transport

After cleaning the video and hi-fi audio heads, clean the parts of the head drum that the tape passes over. Be careful not to touch the heads again. Look carefully for streaks or dirt particles that have adhered to the drum. Use a little extra head cleaner on stubborn spots.

Next, use a cotton-tipped swab designed for cleaning audio heads to clean all the other parts the tape passes over. These include the tape guides, the audio/control head, the fixed erase head, the pinch roller, and the capstan. (For a description of these parts, see Chapter 3, "How a VCR Works.") Don't omit the idlers, which look like plastic wheels with rubber tires. Clean all these parts carefully but thoroughly.

Finally, if there is debris or dirt in other areas of the transport, use a rag to wipe it away. Then that dirt won't find its way onto the video heads. Don't wipe away any grease on the transport; the VCR needs this grease to operate smoothly.

If the VCR has belts visible from the top, clean them with a cotton swab and alcohol. This helps keep the belts from slipping.

After cleaning the video heads, clean the rest of the head drum. Be careful not to touch the video heads.

The belts underneath the VCR could also be cleaned if desired. However, this requires removing the bottom panel (and, if necessary, the front panel), which is usually more difficult than taking the top off. To clean the belts underneath, remove the bottom panel (and the front panel, if necessary) and any circuit boards or sheet metal plates covering the belts. Use numbered cups for the screws, as was done when removing the top. If the belts can be removed, take them off one at a time and soak them thoroughly in alcohol. Wipe them off before replacing them. If the belts cannot be removed, clean them with a cotton swab and alcohol.

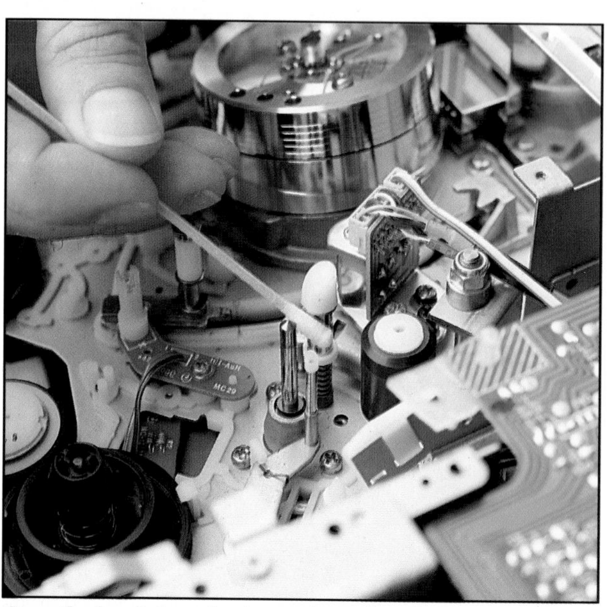

Once the head drum is clean, clean the other parts of the transport. Here, the capstan is being cleaned with a cotton swab.

Once the VCR is cleaned, put it back together carefully, replacing each screw according to its number. Next, temporarily connect the VCR to the system. Try a tape to see if the problems have cleared up.

If cleaning the heads does not clear up the picture, the problem may be a head clog. A *head clog* is a stubborn piece of debris covering a video head. The easiest way to remove a clog is to

Clean any visible belts with a cotton swab and alcohol (top). This will help prevent them from slipping. Some belts may be accessible only from the bottom of the VCR (bottom).

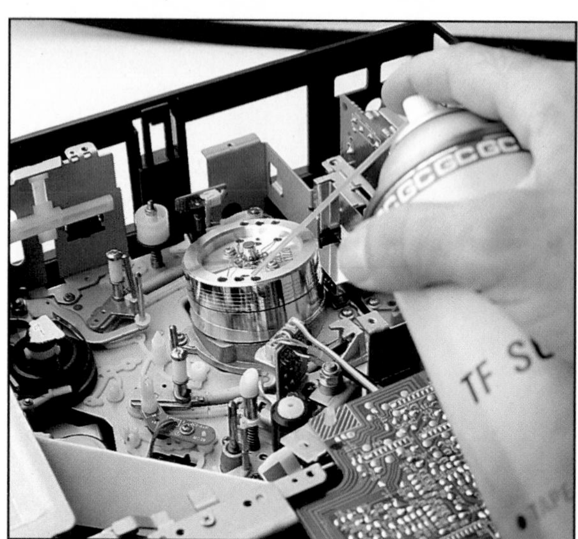

After removing the bottom panel of the VCR, clean these belts with a cotton swab and alcohol (top). To clean a head clog, give each video head a blast of head cleaner from an aerosol can (bottom).

use a can of aerosol head cleaner with a long tube inserted in the spray nozzle. Give each head a direct blast of cleaner for a couple of seconds, then clean each head with the cleaning stick as before.

When Professional Cleaning Is Required

If the VCR develops a picture problem that won't go away with careful cleaning, chances are it's time to take the VCR to the shop. The machine may have an especially stubborn head clog, which a professional repair shop should be able to spot and remove. If the problem goes beyond cleaning, the shop should be able to find it and solve it. That's a much better fate for the VCR than the excessive wear caused by too many cleanings.

How to Store and Care for VCR Tapes

Everyone likes to think they'll be able to watch tapes of their children 20 years from now. Unfortunately, videotape often doesn't last that long. The usual reason a tape doesn't last is poor storage. Improper storage oftentimes can cause the tape's magnetic coating to separate and flake off from the plastic backing tape. Improper storage can also allow dust to get onto the tape. Either of these situations can cause a *dropout*—an interruption of the video and/or audio signals—when the tape is played.

Tapes will last longest in an air-conditioned environment, where temperature and humidity are controlled. It's best to store them inside a cabinet to help prevent dust from settling on them. If tapes are stored on a shelf, use plastic cases for them instead of the cardboard cases they come with. Generally, sealed plastic cases do a better job of keeping dust out. Keep tapes away from heating and cooling vents, air conditioners, direct sunlight, and sources of moisture.

Stack tapes along their edges, whether the long or short edge doesn't matter. Stacking tapes flat, with the reels facing up or down, can cause serious problems. The tape can slip down on the reel, which causes the bottom edge to be curled or crushed. The bottom edge is where the control track resides on VHS and Beta tapes. Wherever the control track is damaged, the tape will probably be completely unplayable, and there is no way to recover what's on it.

Always store tapes on their edges. This will prevent many serious problems from developing.

TROUBLESHOOTING

General Problems

• **Problem:** No picture when trying to play tape

• **Solution:** If the TV screen is black, make sure the TV is turned on. Make sure the VCR is in play mode: The front panel should display either the word "play" or an arrow pointing to the right. Next, check to make sure the correct input is selected on the TV. If in doubt, push the INPUT SELECTOR button or buttons several times until a picture appears. If there still isn't a picture, make sure the cables are connected at the back of the TV and VCR. Wiggle them at each end: If an intermittent picture appears, replace the cables.

If the screen is blue, the connection between the VCR and the TV is probably OK. Make sure the VCR is in play mode.

If there is still no picture, try cleaning the belts as described in Chapter 4, "The Care and Maintenance of a VCR."

To check a belt to see if it is slipping, plug in the VCR with the top off. Don't touch the inside of the VCR when it is plugged in, either with a finger or a metal tool! Electric shocks are a real possibility. Set the VCR on its side, insert a tape, and watch for slipping belts. If a belt is definitely slipping, replace it if possible.

• **Problem:** VCR produces snowy picture on playback

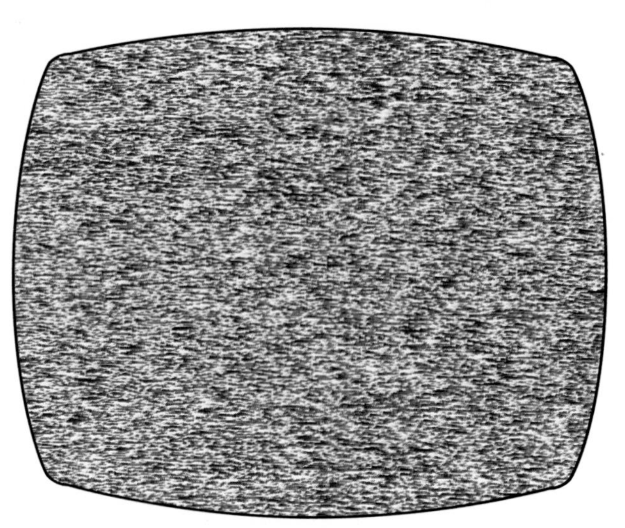

A snowy screen can be caused by numerous problems, most of them easily resolved.

• **Solution:** If the screen is snowy when trying to play a tape, make sure the channel selected (3 or 4) on the TV matches the channel selected on the back of the VCR. Next, try adjusting the tracking control.

If adjusting the tracking control has no effect, make sure the cables are connected at the TV and

Adjusting the tracking on the VCR is one possible solution to a snowy screen.

the VCR. Wiggle them at each end: If an intermittent picture appears, replace them.

If there is some improvement but still some snow, try cleaning the VCR. If the sound is present but the picture is snowy, the video heads need cleaning. Clean your VCR according to the instructions in Chapter 4, "The Care and Maintenance of a VCR."

If the picture still is not cleared up, the tape was probably made on a different VCR, and it most likely will never play correctly on your VCR.

If snow appears when using the tuner on a VCR that doesn't

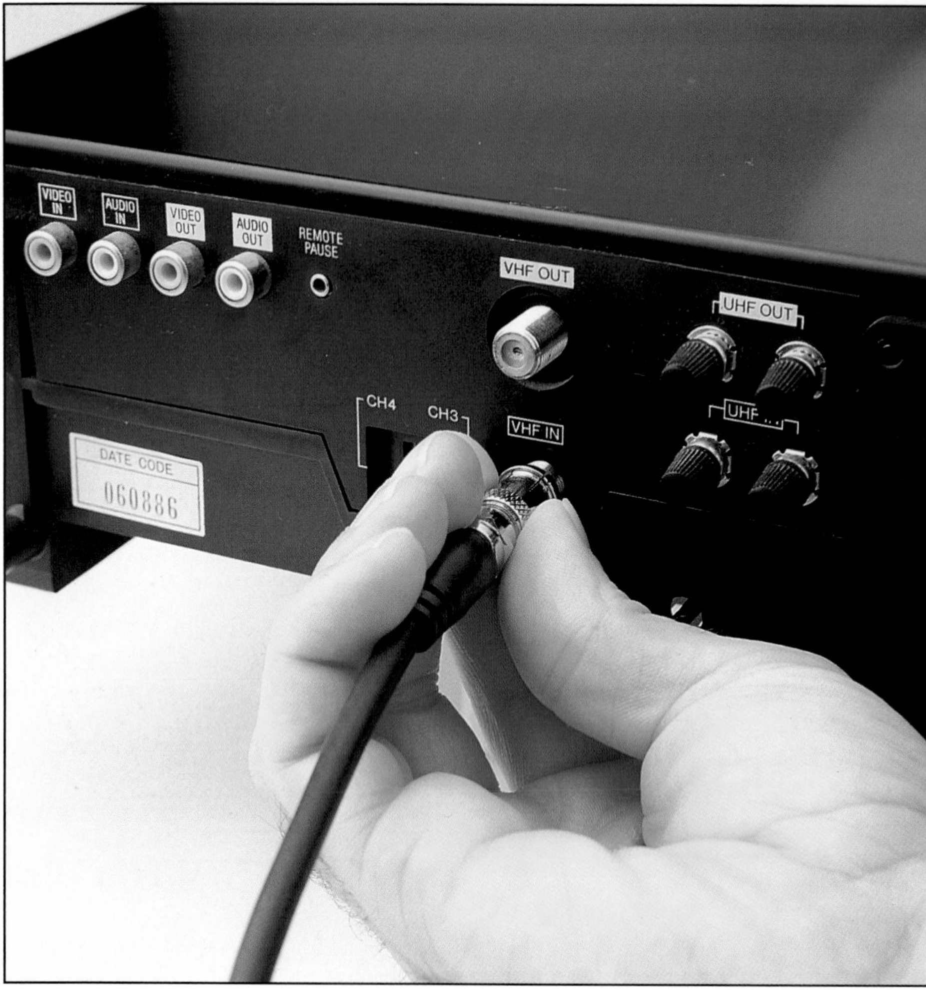

If tracking doesn't improve a snowy screen, try wiggling the connecting cables at the back of the TV and the VCR. New cables may be necessary.

have automatic tuning, make sure the channel is tuned in. If there is another TV set in the house, try connecting the antenna or cable to it. If there is still snow, chances are the cable service is temporarily out or the antenna is misadjusted.

• **Problem:** VCR won't tune in TV channels

• **Solution:** First, check to make sure the cable or antenna is connected to the back of the VCR. (If the cable box is used without a signal splitter, the VCR's tuner won't work in the first place except on the channel the cable box outputs on.) Wiggle the cable: If an intermittent picture appears, replace the cable.

If the VCR does not have automatic tuning, make sure the desired channel is properly tuned in. If the VCR won't tune in channels above 13, make sure the cable/antenna switch is properly set. (Older VCRs that aren't cable-ready won't tune in cable channels above 13.)

• **Problem:** VCR doesn't record shows when user is away

• **Solution:** Usually, this problem occurs when the user fails to turn the VCR off before leaving the house or fails to turn the timer on. On many VCRs, simply programming the timer won't do the whole job. If the VCR was set and programmed correctly, make sure the VCR will tune in the desired channel. If it won't, follow the troubleshooting instructions listed on the previous page, "VCR won't tune in TV channels."

Also check the timer settings, if possible: It's easy to select the

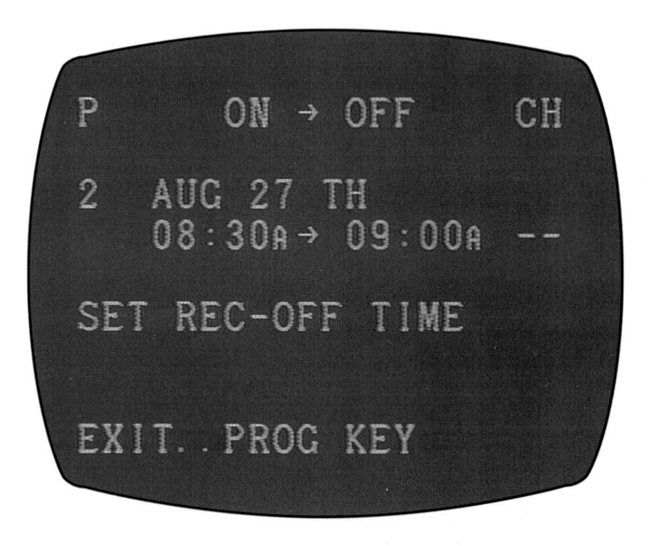

If the VCR won't record when the user is away, make sure the VCR is correctly programmed (top) and the timer is turned on (bottom).

If the VCR won't record at all, the record tab may be missing from the cassette (left). Also, be sure the proper input is selected (bottom).

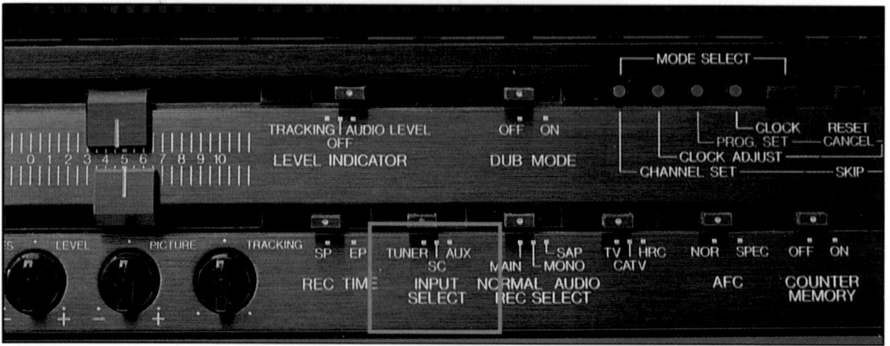

wrong date or channel. Make sure the channel, time, or date wasn't left out when setting the timer. If doubts exist about the timer setting, experiment on programs that won't be taped. Also check the owner's manual again.

With cable, make sure the cable box is turned on and tuned to the proper channel. If the problem still can't be solved, consider getting a VCR Plus programming remote control, which is discussed in Chapter 1, "VCR Formats and Features."

• **Problem:** VCR won't record

• **Solution:** In almost all cases, the reason a VCR won't record is because the record tab has been removed from the tape cassette. In this case, the VCR will accept the tape but refuse to go into record mode. On VHS tapes, the tab is at the rear of the cassette; on Beta tapes, it's on the bottom; and on 8mm tapes, it's a sliding piece of plastic on the rear. If the record tab of a VHS or Beta cassette is punched out, place a piece of tape over the hole. For an 8mm tape, make sure the tab is positioned so that the red part shows.

If the VCR goes into record mode but there is no picture, make sure the proper input is selected on the VCR. If recording a TV program, select the tuner input. If copying a tape from another VCR or a camcorder, make sure the line or auxiliary input is selected. If there is snow when recording a TV program, consult the section above, "VCR won't tune in TV channels."

• **Problem:** VCR won't eject tape

• **Solution:** If the VCR won't eject a tape, the tape may simply be misaligned inside the tape basket. Try carefully nudging the tape loose through the tape hatch with a finger. More than likely, the cover of the VCR will have to be removed to free the tape. For instructions on removing the cover,

If the VCR won't eject a tape, try nudging it loose through the tape hatch.

see Chapter 4, "The Care and Maintenance of a VCR."

Once the cover is off, check to see if the tape itself is caught in the transport. If it is not, try to wiggle the cassette free, being careful not to break the basket in the process. If this doesn't free the cassette, activate the loading mechanism by hand. To do this, look for the mechanism that loads the tape and moves the basket. It may be a worm gear mechanism or a belt drive. Find a shaft, gear, or pulley that is part of the mechanism and gently, but firmly, try to move it. If the basket doesn't move up or out, try moving the part the other way.

If this works, examine the basket for damage and make sure there are no foreign objects jammed in it. If a part is bent, try bending it back. If the deck uses a belt to drive the loading mechanism, clean the belt with alcohol. Now plug the VCR in, being careful not to touch any parts inside in order to avoid getting an electrical shock. Try inserting a tape that can afford to be damaged and see if it works. If it doesn't work, the VCR probably needs a new transport

If nudging doesn't free a stuck tape, move the gear that operates the tape basket.

If the tape doesn't retract completely into the cassette, try moving the gear that operates the guideposts.

loading mechanism or loading mechanism motor.

If a loop of tape hangs out from the cassette door after the cassette ejects, insert another cassette and find the mechanism that moves the small posts that pull the tape from the cassette. Try rotating one of the gears or pulleys for this mechanism. If the guides don't move toward the cassette, rotate the gear or pulley the other way. Don't try pushing on the posts themselves or they may be pushed out of alignment. If the tape retracts into the cassette shell, remove the cassette as described above, and then give the VCR a good cleaning. If none of these procedures works, take the VCR to a repair shop.

• **Problem:** VCR won't load tape

• **Solution:** There are two basic causes for a VCR failing to accept a tape: Either the loading mechanism is damaged or the loading motor isn't working or isn't getting power. Before investigating further, make sure the VCR is plugged in and powered up before putting the tape in. Also make sure no foreign objects are inside the mechanism.

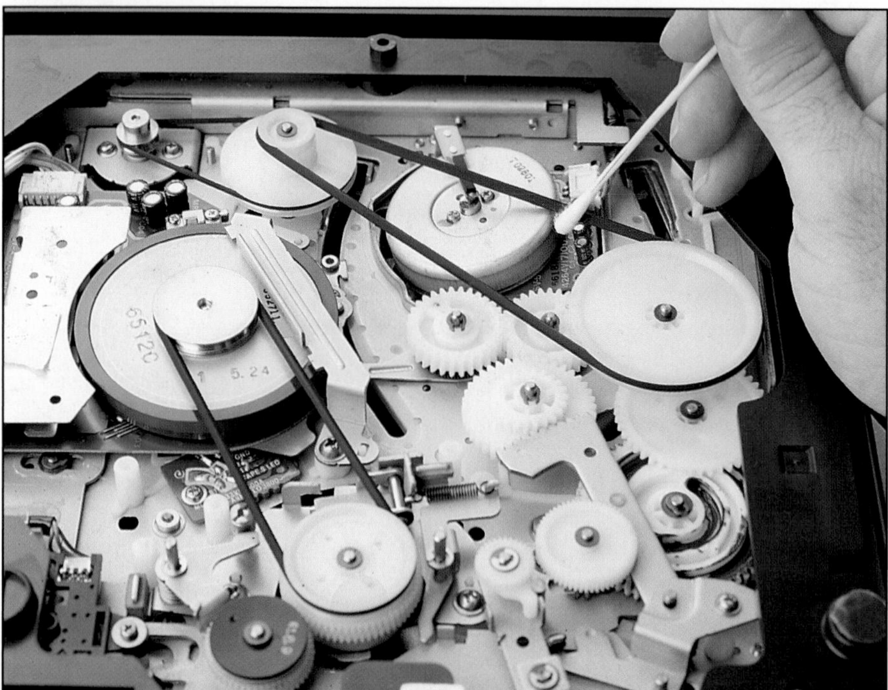

If the VCR won't load a tape, the loading mechanism may need to be looked at. A belt-drive mechanism can usually be found on the bottom of a VCR (top). Clean the belt with a cotton swab and alcohol (bottom).

If the deck still won't accept a tape, unplug it and remove the cover as described in Chapter 4, "The Care and Maintenance of a VCR." First, find the switch that tells the VCR a cassette has been inserted: It should be somewhere on the side of the tape basket. Clean the switch and its contacts by blasting them with head cleaner.

If this doesn't help, inspect the tape loading mechanism to see if it is jammed or misaligned. If it seems jammed, look for the mechanism that loads the tape and moves the basket. It will probably be a worm gear or a belt drive. Find a shaft, gear, or pulley that is part of this mechanism and gently, but firmly, try to move it. If it uses a belt-drive mechanism (which may be located on the bottom of the VCR), clean the belt with a cotton swab dipped in alcohol. If the gear turns, move the basket to its topmost forward position. Plug the VCR in, being careful not to touch internal parts to avoid electric shock, and try inserting a tape again.

If none of these procedures works, the VCR may have a bad motor or gears that have slipped out of proper timing.

• **Problem:** VCR won't power up

• **Solution:** If the VCR won't operate when the power switch is

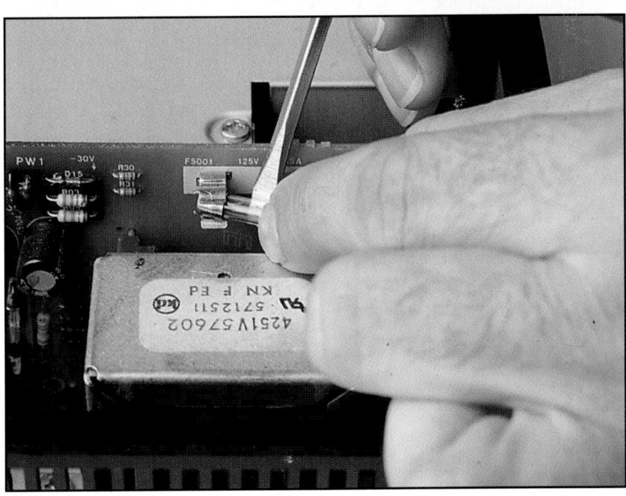

If the VCR won't power up, the fuse (top) may need to be replaced. The fuse usually easily snaps out with a screwdriver (bottom).

turned on, first make sure the VCR is plugged in. If the TV uses a power cord that plugs into the back of the VCR, make sure that cord is plugged in. Check the power cord itself for frays or cuts.

Next, check the wall outlet by plugging in another appliance. If the outlet is controlled by a wall switch, make sure the switch is turned on. If the outlet still won't work, check the residence's fuse box or circuit breakers.

If the VCR is plugged into a power strip, make sure the power strip is turned on. If the power strip has a circuit breaker, try resetting it. Also try plugging another appliance into the power strip to make sure the strip is working.

If the VCR is getting power but it still won't turn on, remove the cover as described in Chapter 4, "The Care and Maintenance of a VCR." Find the fuses, usually small glass tubes. Look inside each tube. A small wire should be visible inside. If that wire is broken or if the glass is blue or smoky, the fuse may need to be replaced. The fuse's value should be printed on the circuit boards in amps (abbreviated as "A") or milliamps (abbreviated as "mA"). For example, a 500 milliamp fuse will have "500 mA" (or ".5A") printed next to it. A local electronics dealer should be able to supply a replacement. To change the fuse, just snap out the old one and put in the new one.

• **Problem:** VCR will not fast-forward or rewind

• **Solution:** The cause is usually a mechanical failure in the trans-

If the VCR won't fast-forward or rewind, the tape sensor bulb (box) may be burned out.

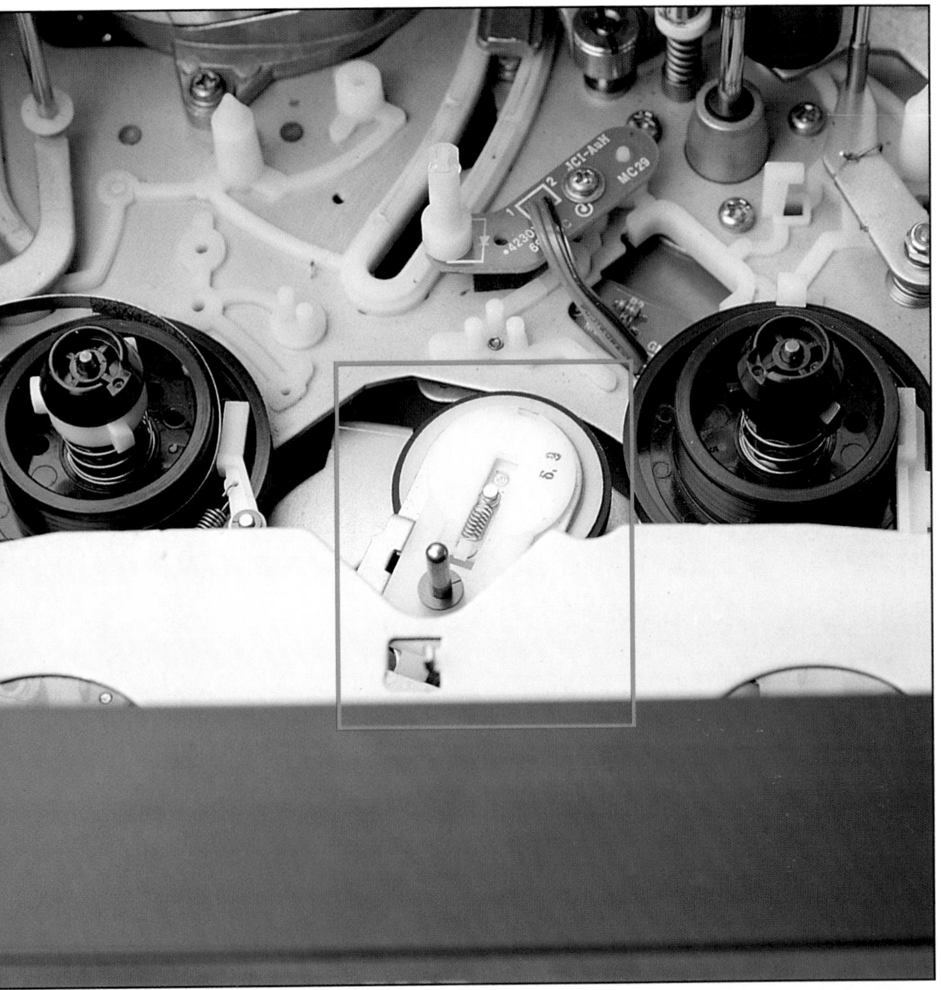

If the VCR won't fast-forward or rewind, and it has no tape sensor bulb, check the idler (box).

port. On older VHS decks, however, the cause may be a burned-out tape-end sensor bulb. In either case, remove the cover of the VCR as described in Chapter 4, "The Care and Maintenance of a VCR." If the tape-end sensor light looks like a tiny incandescent bulb, there's a good chance that it is burned out. Figure out how to remove it—sometimes it just pops out of a socket, and sometimes wires may need to be unplugged. The wires may be soldered into the board. If this is so, take the VCR to a repair shop and have them replace the bulb. Check the bulb by hooking it up to a nine-volt transistor battery. If the bulb is burned out, get a replacement bulb from an authorized repair center for that brand of VCR.

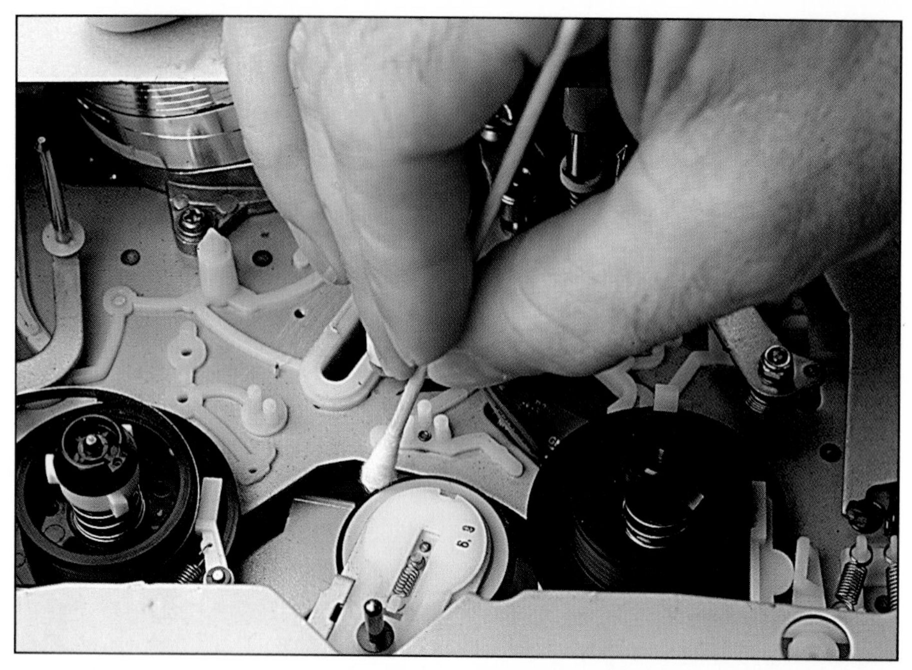

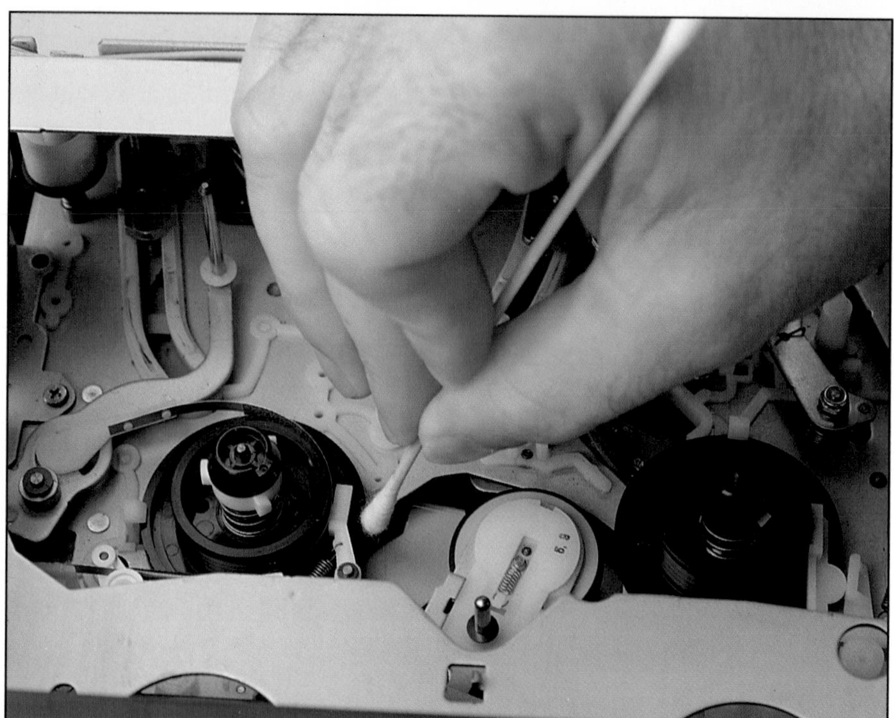

Try cleaning the idler with a cotton swab soaked in alcohol (top). Also be sure to clean anything the idler comes in contact with (bottom).

If the VCR doesn't have an incandescent tape-end sensor (it probably doesn't), try cleaning the idlers, which are small plastic pulleys with rubber tires, and the idlers' contacts. Use cotton swabs soaked in alcohol. The bottom cover may have to be removed to get to some of the idlers and the controls. If this doesn't help, take the deck to the repair shop: It may have a burned out motor.

• **Problem:** Picture appears only when using forward and reverse scan modes

• **Solution:** This problem often occurs only on certain tapes. It is caused by bending or wrinkling of the bottom part of a VHS tape, which carries the control track. In some cases, the control track is so damaged that the VCR can't read it. In the absence of a control track, a VCR won't produce a picture in play mode but sometimes will produce a picture in fast-forward and reverse scan.

Release the tape hatch (on the side of the cassette) and lift it up to examine the tape. If the bottom edge of the tape appears bent, or "cupped," the tape was probably stored incorrectly. Flat storage can cause the tape to slip down on the reel and bend.

If the bottom edge of the tape appears wrinkled or creased, the problem may be a malfunctioning transport. Remove the cover of the VCR as described in Chapter 4, "The Care and Maintenance of a VCR," and examine the capstan, a shiny post that sits next to the rubber pinch roller. If the small, thin

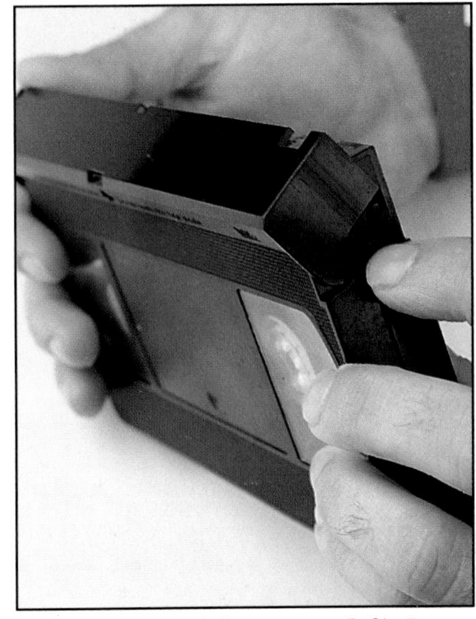

To inspect a tape for damage, look for the release on the side of the cassette (left). Press the release in and lift the hatch (right).

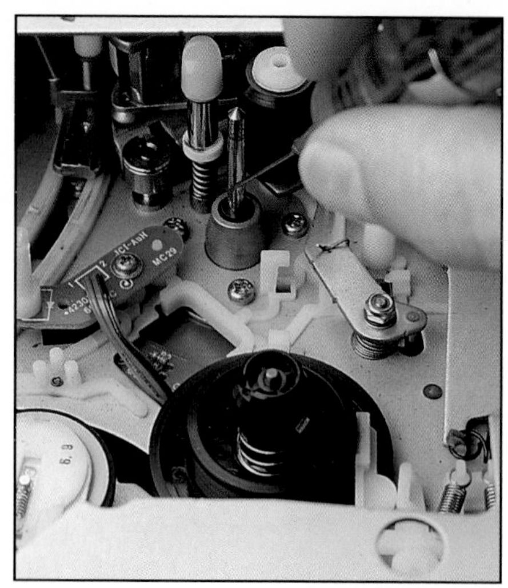

If the tape is wrinkled, examine the capstan (left); the capstan may need oil (right).

bearing around the bottom of the capstan has worked its way up onto the shaft, apply a drop of sewing machine oil at the base of the capstan. Then push the bearing back down firmly. If the bearing is in place, the tape guides of the VCR probably need realignment—a job best performed by a repair shop.

• **Problem:** VCR "eats" tapes

• **Solution:** Eaten tapes are one of the major problems people have with VCRs. Fixing the problem usually isn't difficult. An eaten tape usually stays inside the VCR when the cassette ejects, leaving a loop of tape hanging out from the tape hatch.

First, remove the cover as described in Chapter 4, "The Care

"Eaten" tapes are a major problem for many people.

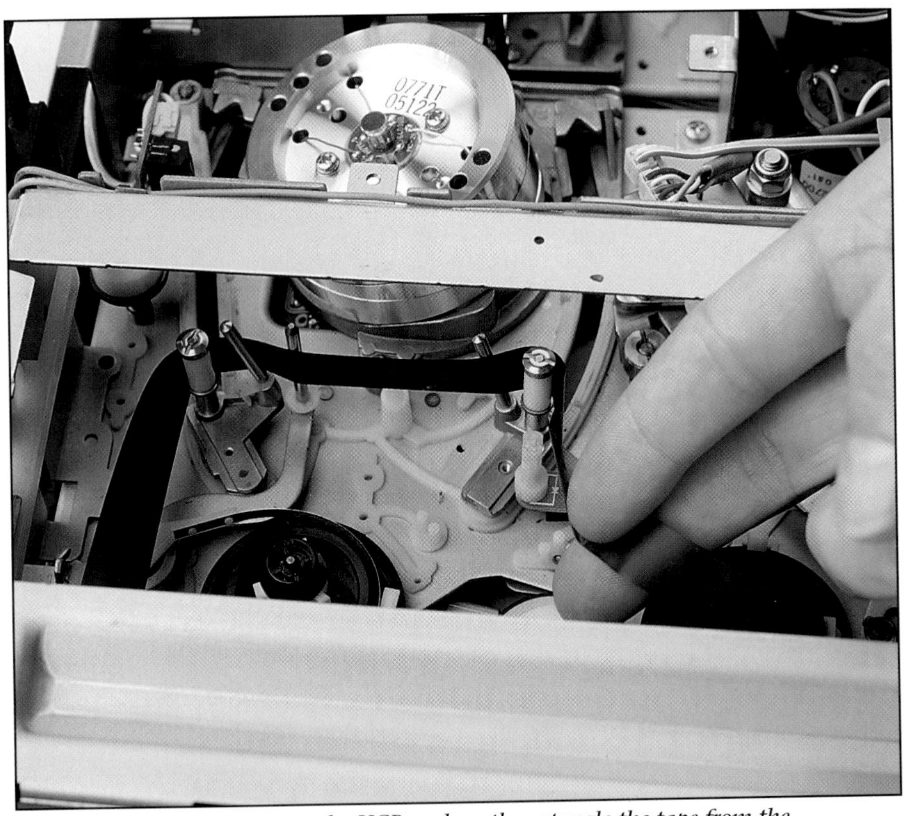

To retrieve "eaten" tape, open the VCR and gently untangle the tape from the guideposts, head drum, or anywhere else it may be caught.

and Maintenance of a VCR." If the tape is tangled around the head drum and/or other parts of the transport, gently untangle it and pull it out from the VCR. Discard the tape. If the tape has personal value to you, try copying it to another tape, then discarding it. Clean the heads and transport as described in Chapter 4.

Eaten tapes can be caused by moisture on the heads, which makes the tape adhere to them. This can occur because the VCR was used right after being brought from a cold environment to a warm, humid one. It can also occur because a wet-type cleaning tape was used to clean the heads and a tape was played immediately afterward.

If neither of these circumstances seems to be the cause, try cleaning the VCR's belts and idlers as described in Chapter 4.

• **Problem:** Tape leader snaps off

• **Solution:** Sometimes a malfunctioning VCR or tape rewinder fails to stop after rewinding a tape and snaps off the tape leader. If the

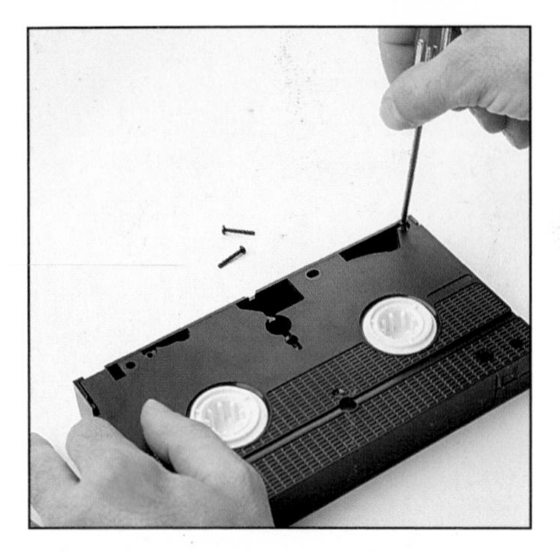

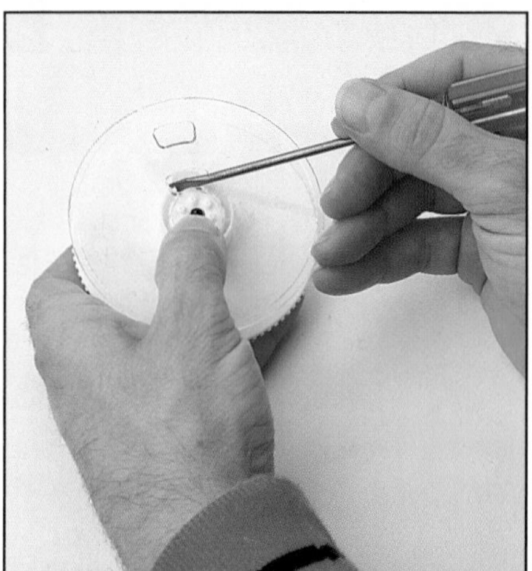

To fix a tape after the leader snaps off, first open the cassette (top left). Next, remove the empty reel where the leader snapped off (top right). Snap out the wedge that holds the leader (bottom left). Finally, put the leader back into its proper slot and replace the wedge (bottom right).

leader has broken near its base, it can easily be reattached. Take the cassette apart by removing the screws underneath. Remove the empty reel and take out the small plastic wedge-shaped piece that locks the leader in place. Slide the leader back into the slot where the wedge was, then put the wedge back in and reassemble the cassette.

If the leader has broken near the tape, it can be reattached using splicing tape, which is available at electronics stores. Or, the tape can be attached straight to the reel, using the same process as attaching a leader. Only use this tape once, to make a copy. Since there's no leader, the VCR won't know when the tape ends, and it may pull the tape loose again. As the

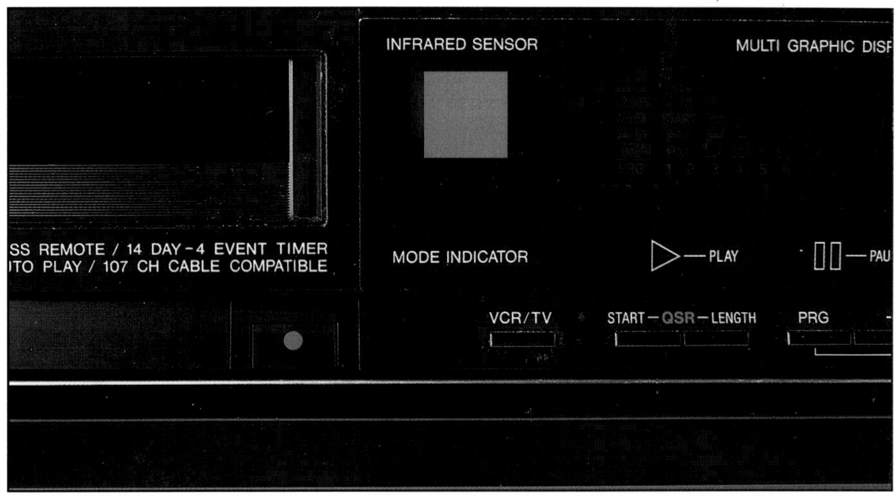

The front plastic window on the VCR for the remote control sensor may need to be cleaned if the remote control isn't working properly.

copying process goes forward, watch the VCR's operation carefully. If the VCR does anything unusual, stop the machine and eject the tape.

• **Problem:** Remote control does not work

• **Solution:** Since remote control batteries often last a year or longer, many people forget they even have batteries. If the remote control works intermittently or not at all, first try cleaning off the plastic window on the front of the remote control. Also clean the plastic window, usually about ½ inch square, on the front of the VCR. If cleaning doesn't work, a new set of batteries almost certainly will.

• **Problem:** Counter seems stuck

• **Solution:** Usually, a stuck counter isn't really stuck: It simply isn't moving because there is noth-

Counters on newer VCRs are real-time counters that depend on the control track of the tape to work.

ing on the tape. VCRs with real-time counters use the control track on the tape to generate the time readout. Since blank spots in a tape don't have a control track, the counter will stop until it encounters another recorded section of tape.

However, in a very old VCR with a mechanical counter (the type that has black wheels with painted white numbers), the belt may be slipping. Remove the VCR's cover as described in Chapter 4, "The Care and Maintenance of a VCR," and clean the belt with a cotton swab soaked in alcohol. If the problem persists, replace the belt.

Video Problems

• **Problem:** Picture filled with lines

• **Solution:** This problem is usually caused by radio-frequency interference, which can come from almost anywhere (radio waves are

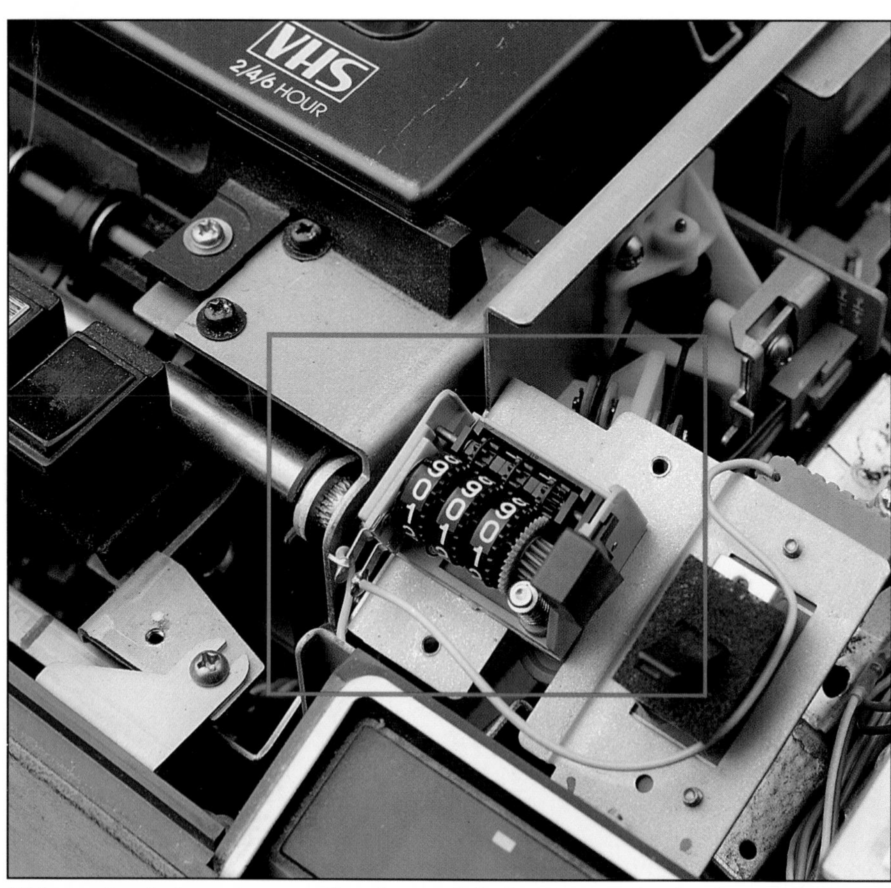

Older mechanical counters usually rely on a belt that may need to be cleaned (box).

Radio-frequency interference is often the cause of a picture filled with lines.

everywhere). To eliminate this problem, move around the cables in the back of the system. Do this with the system turned on, and see if it helps. If the picture cuts in and out as the cables are rearranged, replace them. Better results are also possible by using high-quality cables.

If the TV has audio/video inputs, use them instead of the antenna cable when watching a tape. If the antenna cable must be used, try changing the output channel of the VCR. Turn the channel switch on the VCR to the other channel and find the appropriate channel on the TV.

• **Problem:** Brightness constantly fluctuates

• **Solution:** An annoying fluctuation in brightness may be caused by Macrovision copy protection, which is used to prevent illegal copying of videotapes. If a copy of a movie displays this problem, chances are it's an illegal copy. There is very little that can be done about the problem.

In rare circumstances, original tapes encoded with Macrovision show this problem. If it occurs on a rental tape, take the tape back to the store and ask for another. If it occurs on a purchased tape, take the tape back to the store. Or try playing the purchased tape on a different TV: Some televisions accept Macrovision signals better than others.

• **Problem:** Horizontal line moves down the screen

• **Solution:** A thin line moving down the screen indicates tape damage. The tape may be scratched or may have a deposit of dust or dirt. Usually, there's nothing that can be done: Either suffer with the bad picture or throw out the tape. Sometimes, the line may move to a different spot on the tape. Clean the VCR as described in Chapter 4, "The Care and

There is usually no solution to the problem of a picture with a single line moving down the screen.

help. In most cases, however, there's little or nothing that can be done about the problem, except consulting the repair shop.

• **Problem:** Horizontal snow bands appear on screen

• **Solution:** This is a common problem and is usually easily solved with a quick twist of the tracking control. Turn the tracking control (or hit one of the tracking adjustment buttons). If the picture gets worse, turn it the other way. Continue adjusting the tracking until the picture improves.

Maintenance of a VCR," and make sure the videotapes are properly stored, also described in Chapter 4.

• **Problem:** Picture bends at top of screen

• **Solution:** This problem is usually caused by the TV, not the VCR. In rare cases, a TV just won't accept what the VCR is sending out. It may happen only with a particular tape, or it may happen every time the VCR is used. Try using the VCR with a different TV. Adjusting the VCR tracking control slightly may

Eliminating snow bands on the picture is often simply a matter of adjusting the tracking.

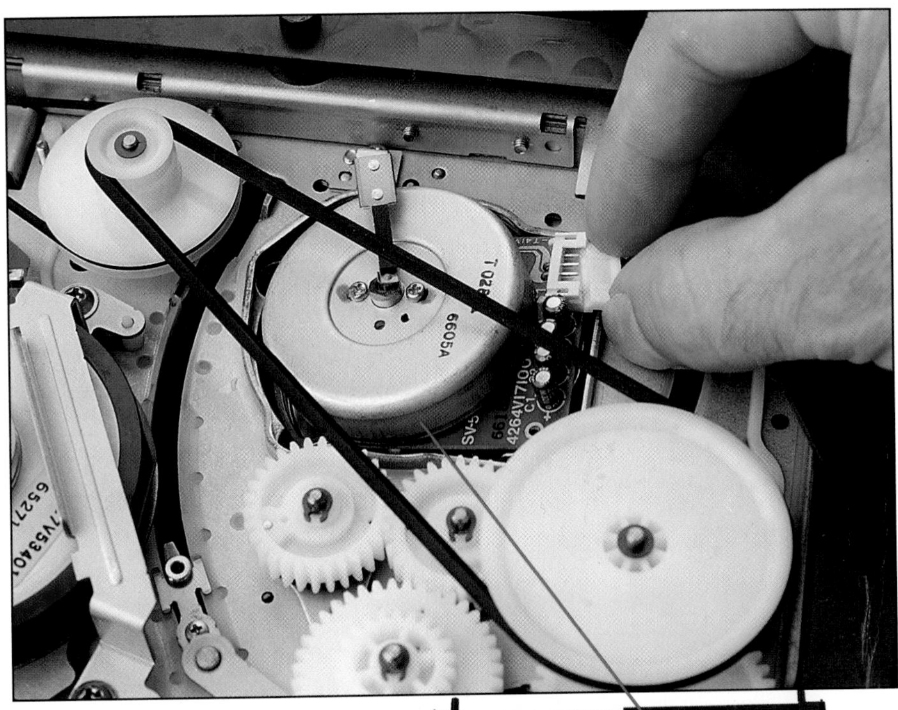

If snow bands persist on the screen, find the wires attached to the head drum (bottom, usually located on the bottom of the VCR). Disconnect and reattach these wires to make a firmer contact (top).

If adjusting the tracking does not help, clean the VCR as described in Chapter 4, "The Care and Maintenance of a VCR." If the problem still doesn't clear up, spray the heads—one of them is probably clogged. Also, disconnect the wires attached to the head drum (they should be attached to a plastic multipin connector on the bottom of the VCR) and reattach them. The wires may have become loose or corroded, and this procedure will allow them to make a firmer contact.

If this snow effect occurs at the beginning of a tape, it can probably be ignored. Depending on the type of VCR used, this effect can occur naturally as much as ten seconds into a tape.

• **Problem:** Rainbow pattern appears across the picture

• **Solution:** This effect occurs at the beginning of almost every tape, as well as between recorded scenes on tapes used to record several TV shows. It's called a *moire* (pronounced mor-AY) pattern and is caused by incomplete erasure of an

The rainbow running down this picture is called a moire *pattern (top). One possible solution to a moire pattern is to disconnect and reconnect the wires attached to the erase head (bottom).*

old program. This is a normal effect on VCRs without flying erase heads. (Flying erase heads are only found on VCRs made for editing.)

If a rainbow pattern covers the entire tape, remove the VCR cover as described in Chapter 4, "The Care and Maintenance of a VCR." Disconnect and reconnect the wires attached to the erase head. These wires may have become loose or corroded, and this procedure will allow them to make a firmer contact.

• **Problem:** Picture jumps horizontally

• **Solution:** If the picture jumps up and down when playing a tape, first try adjusting the VCR's track-

When the picture jumps up and down (top), one possible solution is to check the cassette for any foreign objects. The supply reel is the most likely place to look (bottom).

ing control. If that doesn't help, try another tape. If the problem occurs on only one tape, carefully disassemble the tape by removing the screws underneath. (Refer to the previous section "Tape leader snaps off" for details of disassembling the tape cassette.) Don't touch the tape. Lift the supply reel (the one on the left when looking down from the top) and check for foreign objects that may have found their way into the tape. This problem occurs more often with cheap off-brand tape, which can be of low quality.

• **Problem:** Half of picture is snowy

A picture that is half snowy is often caused by a malfunctioning video head.

• **Solution:** In this case, one of the VCR's video heads is malfunctioning. Clean the video heads as described in Chapter 4, "The Care and Maintenance of a VCR." Assume one video head has a clog and thoroughly spray out all the heads with head cleaner. Unplug and replug the wires attached to the head drum; the wires should be attached to a plastic multipin connector on the bottom of the VCR. (See the above problem "Horizontal snow bands appear on screen" for instructions to do this.) The contacts may have become loose or corroded, and this procedure will allow them to make a firmer contact.

• **Problem:** Glitches or noise appear between scenes

• **Solution:** Glitches can be caused when trying to edit a video with a VCR that does not have a flying erase head. The conventional erase head used on most VCRs doesn't remove old recordings as cleanly as a flying erase head. If many scenes from a camcorder or a second VCR are being edited together, don't hit the stop button on the recording VCR until completely finished. Instead, use the PAUSE button. If the VCR has a flying erase head, glitches are still possible when inserting a new scene in the middle of another

Glitches or noise between scenes often occurs on VCRs without flying erase heads.

scene if the VCR's insert edit feature isn't used.

• **Problem:** Bright parts of picture streak

• **Solution:** Streaking—or overmodulation in technical circles—occurs when trying to play an S-VHS tape on a VHS VCR, a Hi8 tape on an 8mm VCR, or an ED Beta tape on a Beta VCR. Fortunately, this problem is unlikely to occur unless the user owns an S-VHS or Hi8 camcorder. To play the camcorder's tapes on the VCR, use VHS or 8mm tape in the camcorder, or dub the tapes from the camcorder as described in Chapter 2, "How to Hook Up and Use a VCR."

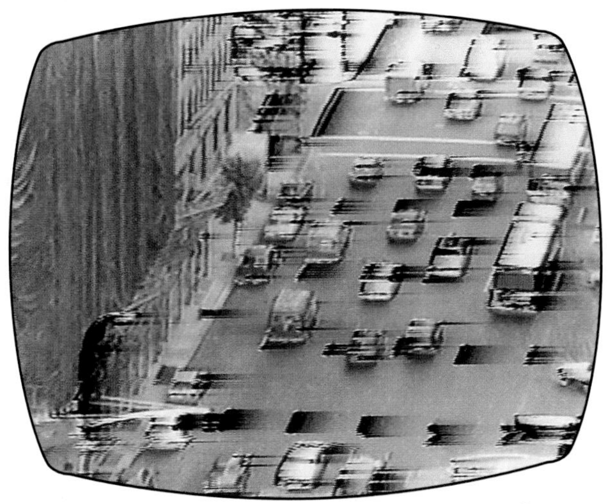

Streaking often occurs when playing a tape in the wrong type of VCR.

Audio Problems

• **Problem:** No sound on playback

• **Solution:** First, check the obvious: Turn the TV volume up and make sure its mute feature (if it has one) is deactivated.

With a hi-fi VCR, try hitting the audio monitor switch. If the tape was recorded from a TV broadcast, make sure the VCR's input control wasn't set to simulcast (the simulcast setting records a picture from the TV tuner and audio from the audio input).

Finally, make sure the audio cables in your system are properly

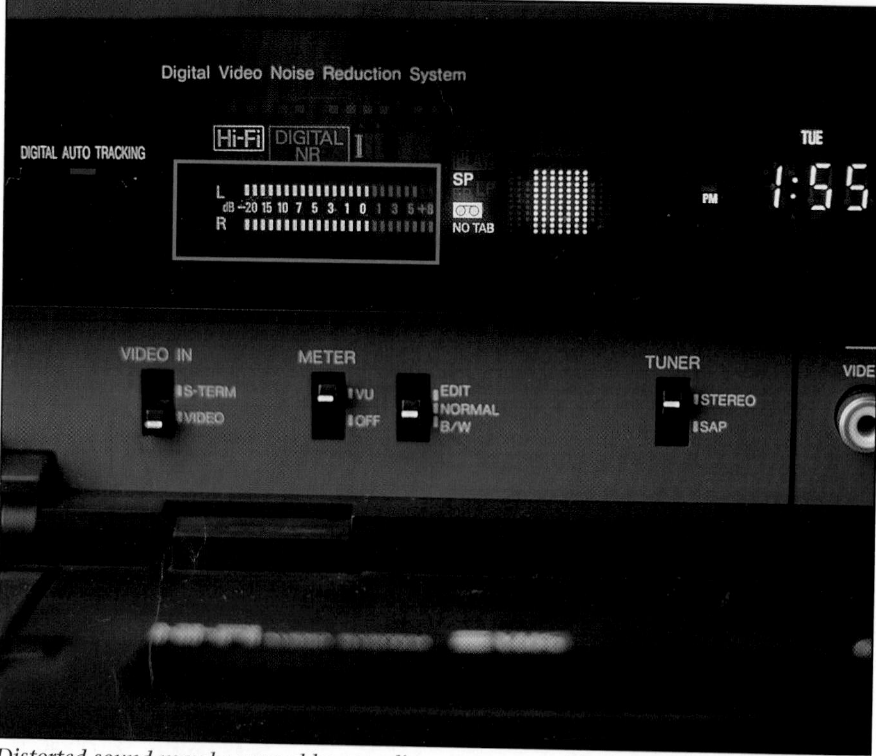

Distorted sound may be caused by recording a program with the audio levels improperly set. The levels should not be set too high.

connected. If they are, try another set of cables.

• **Problem:** Mono sound from hi-fi VCR

• **Solution:** Hit the audio monitor switch and watch the audio track selection indicator on the VCR. The indicator usually takes the form of backlit letters *L* and *R*. For stereo playback, both letters should be lit.

If the program was recorded from a TV broadcast, it may not have been in stereo. Although a program may be supplied to a TV station in stereo, mistakes by the station's engineers or deficiencies in their equipment may cause the program to be aired in mono.

Make sure both audio cables coming from the VCR are connected. If they are, try another set of cables. Also be sure to use the audio/video connections on the VCR; the coaxial output on channels 3 and 4 puts out mono sound.

• **Problem:** Distorted sound

• **Solution:** If the audio on a tape sounds unclear, chances are distortion is causing the problem.

Distortion is usually caused by setting the volume on the TV too high: Most TVs have poor sound systems that can't be turned up all the way without sonic degradation.

If the recording was made using a hi-fi VCR with audio recording level controls, the levels may have been set too high during recording. Turn the levels down slightly and make sure the audio level meters seldom or never peak out. Or set the level controls to the center, which usually activates the VCR's automatic recording level control.

• **Problem:** Crackly or intermittent sound

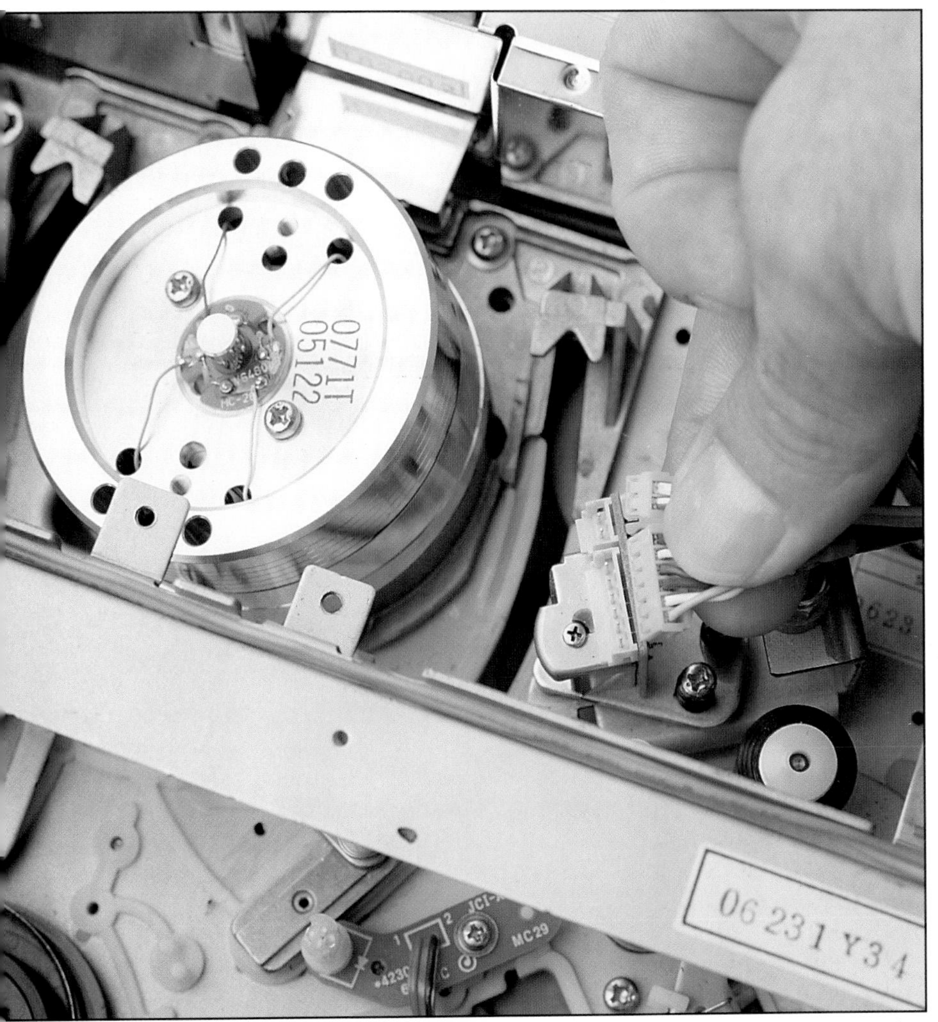

Crackly or intermittent sound can sometimes be fixed by disconnecting and reconnecting the wires attached to the audio/control head.

• **Solution:** This problem almost always occurs because of a bad cable. Wiggle the audio cables coming from the VCR—if the crackling worsens, get new cables. If replacing the cables doesn't help, remove the VCR's cover as described in Chapter 4, "The Care and Maintenance of a VCR." Unplug and replug the cable connected to the audio/control head; in a hi-fi VCR, also unplug and replug the cables coming from the head drum. If none of these procedures helps, the TV or audio system is probably causing the problem.

• **Problem:** VCR won't record sound

• **Solution:** Check the VCR's input switch, making sure it isn't set to simulcast (the simulcast setting records a picture from the TV tuner and audio from the audio input). When dubbing from a camcorder or a second VCR, make sure both cables are connected. If they are, try another set of cables.

When to Take the VCR to the Repair Shop

If the procedures listed above don't solve the problem or if the problem isn't covered here, it's time to take the VCR to the repair shop. If the problem can be fixed, chances are a service center authorized to repair that brand of VCR can do it.

Unfortunately, it won't be cheap. Many people are intimidated by repair shops. They probably should be: There are more than a few shops that are dishonest or incompetent. The best bet is to use an authorized service center. If that shop is inconvenient, ask the dealer who sold the VCR for a recommendation. In any case, know exactly what the problem is before going in and demand an estimate before the shop begins the repair.

The troubleshooting guide above should provide at least an inkling of where the problem lies. The more intelligent the consumer sounds when describing the problem, the less likely the shop will be to try to take advantage.

Before leaving the shop with the repaired VCR, insist on plugging it in to make sure the repair job has been effective. This will save an extra trip to the shop if something's wrong, and the repair shop will know that the problem didn't occur in transit or through further use.

When to Buy a New VCR

Unfortunately, fixing a VCR often costs more than buying a new one. And new VCRs at the price paid years ago usually have new features that did not exist or were prohibitively expensive when the last machine was bought. For example, if an inexpensive VCR needs new video heads—usually a $200 job—it's almost never worth repairing.

Of course, the consumer is the only one who can decide how much money to spend on video equipment. But if the estimated cost of the repair is more than half the cost of a new VCR, it is advisable to drop by a local electronics

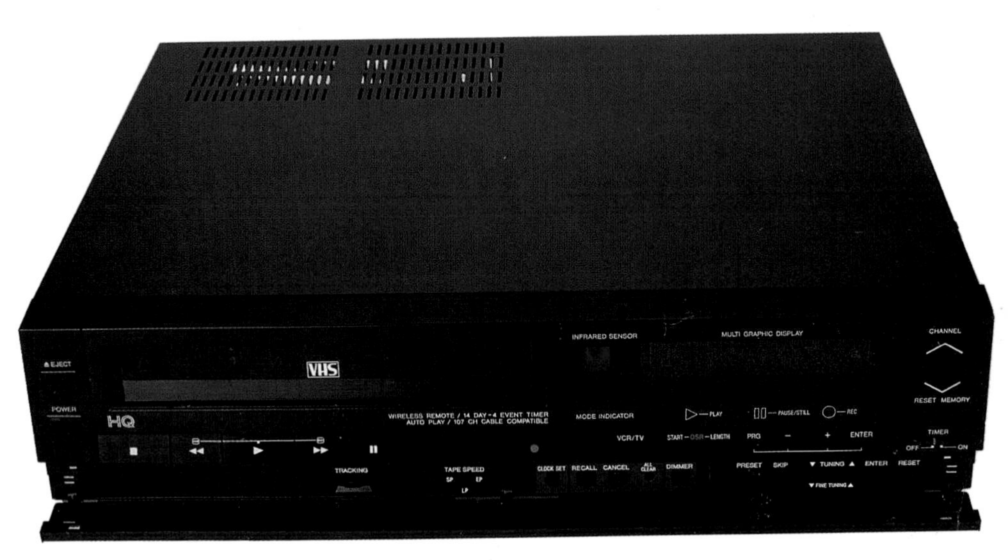

dealer to see what new VCRs can do and what they cost.

A new VCR may be worth buying just to take advantage of new features and capabilities. But that's no reason to junk the old deck. As long as it is well-maintained and clean, the old VCR should serve well as a second VCR either in the bedroom or in the children's room. If the old VCR stops working and can't be easily and inexpensively repaired, it should probably just be thrown away, unless a friend who is an electronics expert and who does not mind engaging in a nothing-to-lose attempt at repair is willing to give it a try.

GLOSSARY

Audio Head: A head used exclusively for reading audio signals. An audio head can be mounted on a head drum or can be stationary.

Balun: A small matching transformer which has screw terminals or wires at one end and an F-connector at the other. A balun converts a cable TV signal into a signal that older non-cable-ready TVs can use.

Baseband: The frequencies at which video can be processed and displayed by a TV set and audio can be processed and reproduced by an amplifier and speakers. The signals from video and audio output jacks are baseband signals.

Beta: The first successful home video format, invented by Sony.

Boost: To increase the voltage of a signal.

Capstan: A small post that turns and pulls the tape through a VCR.

Coaxial Cable: Used for connecting the various component—such as a VCR, TV, cable box, splitter, and so on—of an audio/visual system. A coaxial cable has an insulated wire (or strands of wire) at its center, surrounded by a braid of wire. The braided wire keeps radio signals from interfering with the signal traveling on the inner wire.

Control Head: The device inside a VCR that reads the control track.

Control Track: A thin recording along the bottom of VHS and Beta tapes that consists of pulses that allow the VCR to maintain the correct tape speed.

Deck: Another name for a VCR.

Dropout: An interruption in a video or audio signal due to tape damage. Video dropouts usually look like white lines across a TV screen. Audio dropouts cause a sudden silence or a popping sound.

Eaten Tape: A tape that has been accidentally sucked into a VCR mechanism.

8mm: A small videocassette format used mostly in camcorders. The tape is eight millimeters wide.

Erase Head: The device inside a VCR that erases old recordings from a tape before the VCR records a new program on the tape.

Extended Definition Beta: An improved version of the Beta format.

F-Connector: A threaded input or output jack used to attach antenna and cable TV lines to VCRs and TVs, usually via a coaxial cable. Almost all VCRs today have F-connectors.

Frequency: The number of times something goes through a cycle each second, expressed in hertz. Frequency can be used to describe vibrating objects and electronic signals. For example, when the "A" note in the middle of a piano keyboard is struck, the piano wire must vibrate 440 times per second. Thus, the note has a frequency of 440 hertz.

Head: A device that uses two coils to create a magnetic field when the coils are fed with a signal. The magnetic field changes the arrangement of the magnetic particles on a recording tape.

Head Clog: A piece of dust or dirt on a head that is particularly difficult to remove.

Head Drum: A cylindrical device where the video heads are mounted. The videotape wraps around the head drum during playback and recording.

Helical-Scan: A process that uses heads mounted on a rotating drum to record tracks diagonally across a tape.

Hi8: An improved version of the 8mm format.

Hi-Fi Audio: Audio tracks recorded by the VCR using heads mounted near the video heads on the head drum.

High Resolution: Used to describe a sharper picture. High-resolution VCRs and camcorders are generally regarded as those that produce at least 400 lines of horizontal resolution.

Horizontal Resolution: The maximum number of lines a video device can produce across a screen. The more lines, the sharper the picture will be.

Idler: A plastic wheel with a rubber "tire" used to transfer power from a motor to an axle.

Linear Audio Track: A sound recording that runs along the top of VHS and Beta tapes. It can be mono or stereo.

Noise: Impurities in an electrical signal caused by interference from other signals or electrical devices. Video noise shows up as white streaks or spots on a screen. Color noise takes two forms: A field of color can become unnaturally dark or light in places, or it can show streaks of other colors. Audio noise can be heard as a hissing sound coming from a speaker.

Pinch Roller: A rubber cylinder that presses the videotape against the capstan.

Record Tab: A breakaway plastic tab on a video cassette. If the tab is absent, the tape cannot be recorded.

Reel: A cylindrical object with large flanges on which tape is wound. A take-up reel gathers and holds tape after it has been played or recorded. A supply reel holds the tape before it has been played or recorded and feeds it out as needed by the VCR.

Signal: Fluctuating electrical energy that carries video and audio information.

Snow: Another name for video noise.

Super VHS: An improved version of the VHS format.

Tape Basket: A cage, usually made from sheet metal and plastic, into which a videotape cassette fits inside a VCR.

Tape Loading Mechanism: The motorized device that pulls a cassette into a VCR and positions the cassette so the tape inside the cassette can be played.

Track: The path a head traces along a videotape as it records or plays a signal.

Tracking: The ability of a VCR to read the diagonal video tracks on a tape. Poor tracking shows up as horizontal snow bands in the picture.

Transport: The mechanical parts of a VCR that move the tape around inside the VCR. These include the motor, the tape guides, and the tape reel spindles.

Tuner: An electronic device that selects one channel out of many being broadcast. The tuner then converts that selected signal into one that can be used by either a TV or a VCR.

VCR Plus: A new, simplified circuit for programming VCRs. It was invented by Gemstar and is available as a separate remote control or as a feature built into VCRs.

VHS: The most successful home video format, invented by JVC.

INDEX